The Long-Shadowed Forest

The Fesler-Lampert *Minnesota Heritage Book* Series

This series is published with the generous assistance of the John K. and Elsie Lampert Fesler Fund and David R. and Elizabeth P. Fesler. Its mission is to republish significant out-of-print books that contribute to our understanding and appreciation of Minnesota and the Upper Midwest.

The Long-Shadowed Forest by Helen Hoover

The Gift of the Deer by Helen Hoover

The Singing Wilderness by Sigurd F. Olson

Listening Point by Sigurd F. Olson

The Lonely Land by Sigurd F. Olson

Runes of the North by Sigurd F. Olson

Voyageur Country: The Story of Minnesota's National Park by Robert Treuer

HELEN HOOVER

The Long-Shadowed Forest

ILLUSTRATED BY ADRIAN HOOVER

University of Minnesota Press
MINNEAPOLIS

Copyright 1963 by Helen Hoover and Adrian Hoover, 1991 by
The Ohio University Foundation. Published by arrangement with
W. W. Norton and Company, Inc.

First published in hardcover by
Thomas Y. Crowell Company, Inc., 1963
First University of Minnesota Press edition, 1998

Published by the University of Minnesota Press
111 Third Avenue South, Suite 290
Minneapolis, MN 55401-2520
http://www.upress.umn.edu

A CIP record is available from the Library of Congress.

ISBN 0-8166-3172-7 (pb)

Printed in the United States of America on acid-free paper
The University of Minnesota is an equal-opportunity
educator and employer.

10 09 08 07 06 05 04 03 02   10 9 8 7 6 5 4 3 2

Acknowledgments

THE PEOPLE who help with the writing of a book are a legion that extends across years and spans a gulf from those who encourage with, "Why don't you write it?" to those who challenge with, "What makes you think you can?" I can no more thank these people individually—their number is too large and many of them are shadows from the past—than I can list their tangible and intangible offerings. I make an exception of a little girl, name unknown, who, during a conversation at a nearby lodge, settled a vexatious problem for me. She said, "Boy animals are he, girl animals are she, and if you don't know—it's it." Perhaps she and some of these others may read this and know how grateful I am.

I have special thanks for: Dr. Walter J. Breckenridge, Director of the Minnesota Museum of Natural History; Dr. Clyde M. Christensen, Professor of Plant Pathology, University of Minnesota; Sigurd F. Olson, "Mr. Wilderness" himself, who read and corrected portions of the manuscript; James W. Kimball, formerly Director of the Division of Game and Fish, Milton H. Stenlund and William H. Longley, Game Managers, and Charles Ott, Game Warden, all of the Minnesota Department of Conservation, who supplied current information about fishers, wolves, and deer;

L. P. Neff, Forest Supervisor (Duluth), Superior National Forest, and Ray C. Iverson of that office, who supplied information on the growth and ecology of the forest; John Vosburgh, Editor of *Audubon,* and Richard Cunningham, of the National Audubon Society's Bald Eagle Project, who sent late data on the bald-eagle count; Rev. Bernard Kenny, O.P., and Carl Walther, who mailed clippings that sparked ideas; Florence Mankey and Claire Gomersall, who offered spiritual assistance; and Dorothy Gardiner, whose letters catalyzed a desire to write into an actuality.

For permission to adapt material from some of my articles, I wish to thank *Audubon,* published by the National Audubon Society, *Frontiers,* published by the Academy of Natural Sciences for Philadelphia, *The Living Wilderness,* published by The Wilderness Society, *Canadian Audubon,* published by the Audubon Society of Canada, and *American Mercury.* I also thank *Frontiers* for permission to use the picture of Walter, our adaptable weasel.

Lastly, there are four persons who, beside myself, had most to do with the actual writing, and to whom I am most indebted: Dr. O. S. Pettingill, Jr., Director of the Laboratory of Ornithology, Cornell University, who first suggested that I write this book and who acted as advisory editor; Gorton Carruth, formerly Editor of Reference Books, Thomas Y. Crowell Company, who helped greatly with the planning and who began the editorial work; Edward Tripp, Editor of Reference Books, Thomas Y. Crowell Company, who did the final editing; and my husband, who not only labored into the wee hours on the illustrations but also offered patience and assistance from the beginning to the end of the writing.

HELEN HOOVER

Contents

A Wilderness Trail

A LITTLE TRAIL crosses the hill above our cabins, a barely discernible track, marked in the earth by the four-footed dwellers of the forest as they go silently about their business. It may have been used by Indians long ago, but no human footprints except mine and my husband's have been made there for years. On this trail—or nearby on our bit of virgin woodland that stretches north from the winding access road to border on a lake which laps both Minnesotan and Canadian shores—I have seen most of the things that I will tell you about in this book. I have walked along the track many times, so that every bush and hummock is familiar, yet I am always seeing new things. I never know whether I shall meet several wild neighbors or none. They all use the trail, but when or where I will see them I cannot predict beyond the knowledge that their presence is related to the time of year. Because I am careful to make no smallest alteration, the trail is the place where I can best follow the pattern of the changing seasons. . . .

In late May, after the snow has melted and the ground has dried and the new growth has sprouted, the trail has a dreamlike atmosphere, full of the promise of things to come. Scrolls of pale green are opening on the branch tips of the massive white cedar that hides the trail's entrance. I step past it into the everlasting shadow of the trees, their trunks rising from the springy brownness of the needle-cushioned forest floor like temple columns, filigreed by gray-green lichens and brightened by the emerald velvet of moss. A hundred feet above, the evergreen branches interlace, as pine and spruce, cedar and balsam brace each other against the fury of the winds. Here and there, paper birches, their trunks ragged with bark like shattered armor, show high silvery branches, misty with the purple of buds and the green promise of leaf tips. There is no underbrush in this dim and shaded place, but scattered shoots foretell future low verdure.

Between a pink granite boulder and a rotting stump, I see a spot of white. It is the cup of a calypso orchid, rising on a six-inch stem above its single plantainlike leaf, its five-petaled crown a delicate violet, its cup irregularly striped with red-brown, the whole translucent flower as fragile as foam, immobilized into loveliness. I drop to my knees to breathe its scent, and, while I watch, a minute insect creeps across the yellow pollen-bearing hairs and disappears into the cup.

The path drops sharply into hummocky ground. Here the splash and gurgle of underground water is everywhere and a careless step can sink me knee-deep into a marshy pocket. Pale fiddleheads are cautiously uncurling the fronds on the interrupted ferns. The earth, between hummocks, is carpeted with club moss, whose fine-needled branches rise like miniature trees. And, just beyond, the brook rushes over its stones, pouring the run-off from the spring rains down infant waterfalls as it plays at being a river. At the brook I sit hastily on a stone and remain motionless.

Less than fifty feet away a mother black bear is teaching her baby to lick the tasty mud beside the stream. She sees me and growls. I do not move until she has taken her cub away because mother bears, although ordinarily good-natured, are fiercely protective and very strong.

I cross the brook on natural stepping stones and follow the bank to a patch of marsh marigold, with flowers like splotches of gold that lift above the shallow water they grow in. While I am breathing in the cool fragrance of earth and water, a red squirrel scampers across a fallen log and explodes into furious sound. Twitching his tail, stamping his hind feet, he chatters, moans, squeaks—tells all the forest that an intruder is here.

Leaving him to spread his warning, I climb up a slope where the way is obstructed by the dead lower branches of close-grown young firs, whose slender trunks lift their green tops high, reaching for light. Just beyond, I come to an unexpected clear space several yards across.

Grass and clover are already growing here in the sun, and brown butterflies hover. Sprouts of sweet william and daisy promise summer flowers. In the very center of the clearing is a stand of balsam seedlings, none of them more than three feet tall, a veritable fairy-tale forest that should be inhabited by only the smallest creatures. One of them, a brown deer mouse, crawls timidly from under the little trees to stare at me with wondering big black eyes, and to scurry to safety as an ermine, still wearing patches of winter white in his summer brown, flashes across the clearing into a thicket of alder and red-berried elder. The alder's tassels hang above its tiny cones, and the elder is already heavy with leaves and budding its white flower clusters. I follow after the ermine and pick my way through the brush.

There is a fluttering in the low branches and I look up into the inquiring eyes of a gray jay. A friendly bird dressed neatly in shades of decent gray and black, it flits ahead of me, pausing now and then to whistle softly, much like a gentle old lady who is perpetually surprised by every happening around her. At the far side of the thicket, the jay flies high into the pale-leaved branches of a grove of young aspen trees.

Here the sun shines brightly through the still-sparse leaves and the warm ground between the slender gray trunks is covered with the opening buds of wild strawberries and the four-bracted stars of the bunchberries, just starting to change from green to white. There is a rustling at one side, and I catch a glimpse of a brown snowshoe hare, hopping from me in terror. Or perhaps he hears, as I do, the bark of a red fox nearby.

I stay a long time in the aspen grove, idly shooing away some hungry mosquitoes, because beyond it the path crosses our little road and goes into a cedar swamp, a place danger-ous of footing when the ground water is high, so that it would be foolhardy for me to enter it alone without first telling Ade where I was going. Lovely as the scene looks to me, standing amid the sun-dappled flowers and listening to the song of a robin, I must not forget that I am in the midst of primeval forest.

I decide to follow the road home and look for raspberry canes, but first I must investigate a trampled patch of green-ery near the thicket I have left. Deep-pressed through the leaves and into the moist earth is the track of a moose. Per-haps he has been watching me, this wide-antlered, hump-shouldered giant, and then has slipped away through the thick brush with his amazing and characteristic silence.

Along the road I pause to admire a fat, brown toad, not long out of hibernation and drowsing in leafy shade. The pin cherry trees are white with bloom and—I stop.

Watching me with doubtful, gentle eyes is a doe, her new-born spotted twins wobbling on unsteady legs at her side. She does not herd her fawns into the brush but, after in-specting me carefully and perhaps remembering that I had fed her during the past winter, she decides that the road is safe. And so, as I have been following the animals' trail through the forest, the doe leads her children along the smooth way that man has made.

The trail has a feeling of urgency about it in August, when the summer is at its height and the time of the frosts is nearer than you think. It is as though the plants knew that they must hurry to ripen their seeds.

I stand in the sun, which burns hot and bright in a glaring sky, and regard the cedar tree. Its branch tips are heavy with bunches of immature cones—little flowers whose petals, not yet separated, look as though they had been molded from dull-yellow wax. Beyond the cedar, the air is cool with the moisture breathed from the leaves of the big trees and with the shadows they have cast for more than a century. The evergreens have reached the size at which they show no change, from beneath, to the casual eye, but the birch branches now are thick with bright-green leaves, and, splintered and fallen in the wind, a great black spruce lies across the trail ahead of me. Wild sarsaparilla raises flat nosegays of ripening blue berries and clusters of leaves like torn umbrellas. The pale-yellow lilies of the clintonia have come and gone, to be replaced by metallic blue beads on tall stalks. The tiny purple-pink bells have fallen from the stems of the twisted-stalk and a striped chipmunk dashes up to pick one of its waxy red berries. He sits up, turning the berry over and over in his thumbless hands, eating it as though he were a child with a big red apple.

With a rasp of stiff wings and a cackling, a pileated woodpecker flashes down to the spruce stump, fluffs the sleek black feathers of his wings and back, and cocks his head. I lean against a pine bole and enjoy his pointed scarlet crest and his beautifully marked black-and-white head and throat.

He whacks at the stump and chips fly as he excavates to the insects within.

When he has completed his meal and flown away, I go down the slope to the land that is marshy in spring. Now it is all little hills and valleys, with its club moss almost buried under a flourishing blanket of large-leaved aster, from which flower stalks lift unopened buds. The fronds of the interrupted ferns are tall as my head, their central spore-bearing leaflets brown and withered. The brook is not very busy and bubbles gently within its rocky channel.

I look for tracks in the mud of the lick. A mink has been there, but nothing else. The bears are in the hills eating blueberries, the deer are grazing in seclusion, and many smaller animals prefer puddles for drinking ponds. A cloud of late black flies suddenly hums around my head and I hurry across the brook and through the thicket of firs, where curling feathers of dull-green moss carpet the ground and the scarlet spore caps of the British soldier decorate a rotting log.

Grass is knee-high at the edge of the clearing and the open space is covered by a tapestry of flowers, brought to life by clinging bumblebees and flies. Yarrow and daisies and pearly everlasting make a background of white, with variegated green leaves. Tall buttercups bring patterns of yellow, and sweet williams, of every shade of pink and red, of patterned white and deepest purple, fill in the empty spaces. The little evergreens in the center are half buried in flowers. This is much too beautiful to disturb, so I skirt the meadow's edge to reach the brush beyond.

New green cones are forming on the alder, and the elders bear heavy bunches of red berries. There is a bird song, so clear and high and pure that it stops my breath, and a hermit thrush comes to feed on the elderberries. It goes on to a clump of thimbleberries, where it is hidden by the big notched leaves as it plucks ripe fruits that look like raspberries but are large enough to fit over my fingertips.

The sun dapples the aspen grove, where warblers whisk through the foliage. The violets and strawberries stopped blooming long ago and are hidden beneath two-foot dandelion leaves. The bunchberry carpet now is patterned by clusters of half-formed little green fruits. A breeze from the south brings me the damp, generic smell of the wetland across the road and, from not far away, a musical bellowing. The moose are in their swamp.

The little road is dusty and hot. A jumping mouse leaps across it, looking like a small brown frog. It disappears into the honeysuckle bushes, where golden butterflies flutter over the orange-and-yellow flowers. I nibble a few raspberries that the squirrels and chipmunks have not found. A horn toots, harshly out of place.

I pause at the roadside until the car has gone on in a gritty cloud and a miasma of gasoline. Then I turn off the road and scramble downhill through the brush to the little trail. I follow it back the way I came, marveling at every step that it can exist, so near to the passage of modernity, and yet so completely untouched.

Harvest time comes to my trail in October. The big cedar that guards the entrance is dappled with brown leaves among its green. A red squirrel is rushing from branch to branch, cutting down the clusters of ripened cones, each a perfect little wooden blossom. He watches anxiously as I pass, as though he fears that I might rob him of his laboriously collected winter food.

The shadowy ground beneath the pines is crisscrossed with the rusty needles selected for discard this season. The birches are already bare and the slow decay of the fallen deciduous vegetation fills the air with the friendly scent of leaf fires in small-town streets. Long, fanned rays of light, glittering with dust, reach from forest ceiling to floor. The sarsaparilla greenery has faded to pale buff and most of its blue berries are gone. At the foot of a pine, late Indian pipes are curling gracefully, pallid ghosts of plants. Under a loose pile of brush, I see mushrooms, tiny white buttons with rosy top centers, beige-lined giants like elfin dining tables. The squirrels have been busy harvesting, but there are more than enough for them and for Ade and me, too. I will come back later with a pail. The patch is so large that I can gather all we need for several days.

The hummocky slope is treacherous again, partly from the slippery fallen leaves and partly from the dagger-sharp bare spines of the big ferns. The large-leaved asters are still green in spite of our frosts; their sparsely petaled blue flowers have bloomed and faded into cottony tufts of seeds. The brook is silent, flowing sluggishly from one shallow, silt-bottomed pool to another, sometimes disappearing underground.

The mud lick is dried out, but it holds the forepaw print of a bear—and a track like a broken heart that marks a deer's passing. A wind sighs through the evergreens and carries the faint, faraway honking of wild geese.

I hurry through the fir thicket into the clearing so that I can see the sky. It is the sharp, bright, deep blue that belongs peculiarly to autumn skies in the north. Across its cloud-lessness, the geese fly toward the south, three wedges of them, their strong wings defying the wind and their voices calling from a freedom no man can ever know. Listening to their fading cries, I wish them a safe journey.

One of the little balsam trees in the clearing's center is grown up, in spite of its small size. It has two purple cones on its top, like well-earned decorations. All around me are deep-blue wild asters. I count eighty-seven small fringed flowers on one of their natural bouquets. There are a few late daisies and, like a splash of sunset, one red sweet william beside the clump of evergreens.

The thicket beyond the clearing is leafless. The elders have set tight buds, ready for next spring's early opening. I watch a colony of black ants, coming and going in single file from a hole in an alder stem. A chipmunk appears from his doorway under a stone, snatches an ant, and settles himself on top of his lintel to eat it.

Beyond the thicket, the aspen grove is roofed over with flaming orange. I might be standing in a bed of giant chrysanthemums. Dark against the pale aspen trunks stands a young blue spruce. Its branches range from tip to ground in perfect, graceful drapes and it is trimmed all over with the golden disks falling from the aspens. As I turn toward the road, a ruffed grouse, feeding on scarlet bunchberries, raises his ruff and retreats a few pigeon-toed steps. When I move away, he returns across the bronze leaves of the berry plants to feed again.

A squawking comes from the swamp and I cross the road to look beyond the murky waters out of which the cedars rise, thick of trunk and very old. A flock of crows have settled on one of the trees, flopping from branch to branch, noisily cawing back and forth. They see me and rise, like jagged, misplaced bits of night, then settle again on the same tree to resume their crow conversation, perhaps a discussion of the merits of leaving for warmer climes.

The roadsides are walled with brown and yellow, and, in a protected spot, a striped maple is still covered with rose-red leaves, gorgeous against a background of dark needles. There is movement ahead of me—a wide, bristly, frosted-black rear, swaying from side to side with its owner's plodding progress. A porcupine; a big, old, fat porky! He highlights my walk because porcupines are rare here and I had not known that any lived in my part of the woods.

His pace fits the fulfilled mood of the evening, so, matching my speed to his, I follow him down the road as the sunset copper-plates the spruce tops.

In January, when the big snows have buried the ground and frozen onto the trees, it would not be easy to find the trail after I pass the cedar at its entrance if the deer had not gone over it before me. Stepping, and stepping again in the same footprints, brushing a valley with their bodies as they move through the four-foot white layer, they have marked the way for my snowshoes.

Although the evergreen boughs are thickened and fused together by the snow blanket, colorless sunlight filters through the bare upper branches of the birches. The shadows of the big trees are pale on the snow, which reflects light from every crystal. It is brighter in this cloistered place now than at any other time of the year. There is no sign of the earth beneath the smooth white layer, from which the tree trunks rise rootlessly. Nothing seems quite real.

Fifty yards ahead the tame deer have made a short side trail and are watching me—a buck, a doe, and a half-grown fawn, standing in a little yard they have trampled around a growth of maple brush. I empty cracked corn from a sack, in piles along my trail. As I reach the top of the hummocky slope, I look back.

The doe and fawn are eating the corn, nose to nose. The buck is keeping a properly watchful eye on me. Reassured, he nudges the fawn away and joins the doe at the feast, while the fawn hurries from one pile to the next, hastily licking the top from each one. They finish the last grains, switch their tails, and, in their constant, winter-driven search for browse, go away from me along the trail.

A snowshoe hare has thumped over the white smoothness that hides the hummocks and hollows, leaving prints so sharp that they show the marks of claws on the hind feet. A trickle of snow falls from the edge of one of the tracks. The hare has just passed, probably while I looked the other way

at the deer. The lacy trail of deer mouse ends in a flurry
and a drop of blood, where the marks of an owl's wings lie
like fans. A fox has struggled shoulder deep, leaving neat,
round holes punched by dainty feet. A wide, flat sheet of
ice curves along the stream bed. From under the layers of
ice and snow comes the faintest sound of murmuring water.
The drifted rise that leads to the firs is hard going. I dis-
lodge a miniature blizzard from the twigs as I struggle
through the thicket. I emerge in the clearing, brushing snow
from my eyes.

The open space is sparkling in the sun, its central stand
of little balsams entirely covered and turned into a Lilli-
putian mountain of crystal. Brown- and black-capped chickadees, fluffed into feather balls, are flitting from branch to
branch among the alders. One flies to my shoulder. *"Dee-dee-
dee,"* he says. *"Dee-dee-dee!"* I crumble graham cracker in my
hand. He lights on my palm, selects the largest piece, and
flies away, dipping and rising under the weight in his small
beak. I scatter the rest of the crumbs on the snow and the
whole flock gathers at my feet, the busiest and liveliest bits
of bird life in the forest.

As I pick my way past the cotton-trimmed elder stems, I
hear a doleful croak overhead. Through the white lattice of
branches I watch a raven flap by.

The low sun sends the shadows of the aspen trunks in
blue bars across the white ripples in the grove. There are
deep wind rings around the trees, and a steady current of
air is moving the sugary snow, lifting it lightly, carving it
gently into drapes and frozen waves. With a flash of sapphire and turquoise, a family of blue jays perches above me.

They know me and set up no alarm. As they call and whistle coaxingly, I pour out the last of my corn. They cock their crested heads, watching me as I walk toward the road. They will fly down for the corn as soon as I am well out of the way.

I have just come to the big ridge thrown up along the road by previous passages of the snowplow when the jays set up a frantic screaming and the chatter of red squirrels breaks out behind me. A fisher bounds past me, a black looping streak, incomparably graceful. He disappears across the ice of the frozen swamp, with the jays flying and calling above and ahead of him, warning all the small wild things that a mighty hunter is abroad. Soon the squirrels are quiet and the jays return to argue with them over the corn. I climb across the snow ridge and pause at the road edge.

The road leads south to the world that my husband, Ade, and I left nine years ago to turn our faces northward toward the virgin forest. We have lived here ever since, though most of our few summer neighbors go south with the birds when the autumn chill begins to settle in. We merely close up our frame summer house and move into our snug log cabin to await eagerly the changes that winter will bring. Now, starting down the white road, I ask myself what we are doing here, so far from the busy, often engrossing life we used to lead in Chicago. The answer is not long in coming. Nowhere could Ade and I be more deeply involved with the world around us. There is hardly time enough to appreciate the subtly changing seasons, to watch or share the experiences of our animal neighbors, to enjoy to the fullest the infinitely varied life that surrounds and includes our cabins. The long white road that lies so near brings us pleasant memories, and no regrets.

The plow has not yet come after yesterday's snowfall and the way lies ahead of me so quietly unbroken that I hesitate to walk along it. The scent of the forest is frozen. Its sounds are muffled. The trees bend under their snow burdens and bridge the road with coruscating white rainbows. Under the snow, the life of the forest waits. Here, where winter holds all the land in a white-velvet grip, is peace—which, like silence, is everywhere until it is broken.

King Weather

ON A MAY MORNING some years ago, I was talking with Awbutch, the daughter of Netowance, both of whom are Chippewas, born and reared in the Canadian forest across the lake. It was a lovely day, soft with the earth-perfumed wind, tender with purple birch buds and new grass. Robins sang, and even the raven that spends the winter croaking dolefully around our cabin was making pleasant coaxing sounds for the enticement of a lady raven, sitting aloof in a nearby tree.

"Mother says snow—big snow," Awbutch mentioned, looking up at the clear blue, where puffy clouds like gold-edged roses drifted. "Did you see the northern lights last night?"

I said that I had, and very beautiful they were—mist-green banners tipped with pink.

"I never heard that lights in the spring meant snow. Did you?" Awbutch asked. When I shook my head she said, "Well—Mother says snow." And we both laughed.

The sun went down in a red sky, the "sailor's delight" kind of sunset. Netowance's snow looked very doubtful.

In the morning the calling of blue jays woke me. I looked out through a curtain of feathery flakes so thick that I could

hardly see the trees. The raspberry canes were buried and only the top half of a six-foot elder bush was visible, its new leaves emerging in a surprised way from the snowballs that surrounded them.

The yard was full of birds. Residents that fed with us all winter—chickadees and nuthatches, blue and gray jays, hairy and downy woodpeckers—were calling from the branches, perching on the woodshed roof, peeking in the windows, confident that breakfast would be forthcoming. There were summer birds, too—robins and thrushes spar- rows and warblers and goldfinches—that had followed the lead of the others and were hopping over the drifts in a hope- less search for food. The whiteness of the forest would be dotted with starved and frozen feathered bodies before a thaw cleared the ground. But I could help these few, at least. They could stand the cold if they were fed.

Hastily I broke up graham crackers and crumbled suet, which Ade scattered outside along with seeds and cracked corn. I added canned peas and cherries to the menu and was mixing cornmeal and peanut butter, which robins some- times will take from feeders, when Awbutch pulled her boat in to the shore and plowed to the house on snowshoes that sank deeply at every step.

"Old bread," she said, handing me a parcel. "To trade for corn. So many birds—they'll die."

When she had pulled her loaded packsack onto her back, we stepped out into the storm. She slipped her feet into the thong hitches of her snowshoes and said, with the infinite contempt of one who has a personal weather prophet, "The *radio* says 'snow flurries!'"

I can never hope to equal Netowance's skill at "feeling" the weather, but I wish I could, because weather is the in- disputable ruler of the forest and every living thing in it, man not excepted. Ade and I gear our lives to its changes, and watch eagerly for the dramas that it brings.

The enormously vital spring ice break-out begins with the first thawing day, when the expanse of snow that glares in the sun from shore to shore loses some of its reflecting power. Gradually it becomes grayish and pools of water lie on the ice, mirroring the sky as pale blue or the cloud cover as dull silver. A fan of silt-brown water spreads from the mouth of the brook. In the distance, for the first time since the previous fall, I hear the faint, skirling cries of the herring gulls, piping in returning summer life.

The leaden ice sheet begins to rot, honeycombing from above and below, until it looks gray-black and sags toward the water under it. Our winter water hole floods and, all over the lake, water seeps up through thin cracks to spread in sheets on the ice surface, freezing at night and thawing in the day.

As the ice deteriorates we look along the shore for cracks. Eventually we see a rivulet of water there, deeper blue under the open sky than the ice-bottomed pools, or more mirror-like under the clouds, or a richer copper at sunset time. Soon the ice cracks away from the shore and a pair of loons rests on the open water before they fly on, leaving the echo of their calls. More cracks widen, and the network of water gleams violet and indigo against the mat-gray background. As the sun sets, the water turns from gold and copper to lemon color, then to faint green before it whitens and loses the light and is one with the night's dimness.

One day the wind rises. The ice rumbles, snaps, breaks into floes, and begins to move. It may reach shore in huge plates or it may grind itself into fragments first, but its power is inexorable. Poorly anchored docks are lifted and dropped back as piles of planks and logs. Boulders as large as automobiles are maneuvered into new positions. On the windward side of the lake, open water stretches out from shore. Then, with a suddenness that never loses its exhilaration, the last fragments of the winter freeze-up grind themselves into crystals and tinkle into nothingness. The blue waters advance sparkling in their wake; gulls wheel and cry above the dancing wavelets; a fish flops in the midst of spreading rings. The land may be deep under snow or spattered in any degree with its melting remnants, but the burgeoning time has come back to the forest.

It is only a half-dozen weeks until the longest day of the year, when false dawn dimly lights the clearing at 2:30 in the morning, and the long twilight does not vanish from the west until after ten the next night. The sun rises far to the northeast, swings around a high arc like a drawn bow, and disappears behind the northwestern hills.

The green-and-blue days of the North Woods summer are familiar, not only to vacationers, but to anyone who watches television or goes to the movies or reads magazines or books. Fewer people know the very early dawns.

The first light reveals the trees and brush as greenish-gray shapes, like amorphous, primordial predecessors to themselves. Across the lake, mist rises and coils above a marsh, ghostly and smoke-white as the light increases and the green growth loses its formless mystery. Somewhere a bird chirps, a squirrel sputters into lively argument with itself, gulls pass like shadows overhead, and the eastern sky flushes with pink. The rim of the sun lifts over a hill to send a sparkling path across the lake, and the dawn wind rolls the water gently in bands of rose and powder blue.

From many sunsets watched on warm, moist evenings come a few to remember all your days. Ade and I once saw a thunderhead rise in front of the sinking sun, the cloud's face black and ominous and streaked with blue lightning, its top, boiling into the brilliant upper air, rimmed with fire. Slowly it moved eastward, gray rain veiling the hills beneath it. As it passed north of us, rain swept below the still

rising mass in violet draperies thousands of feet high, sway-ing and folding as the thundering cloud pushed steadily ahead, its bottom in shadow, its sides dull red, and its top turned to molten metal by the upward slanting rays of the sun. Ahead of the storm, the sky was pale green and the hills were indigo. Our lake lay quietly unaffected, turning slowly from mauve to gray. Behind the cloud, a rainbow formed, with a soft ghost of itself below. This double bridge followed the storm until darkness dissolved it.

The strength of the forest is measured in water, and its safety—and ours—depends on rain. Our only frightening days come in a spring when the land is clear of snow and dry before the ice goes out. The dehydrated trees and brush, not yet moistened by rising sap, and the wind-dried duff, are invitations to forest fire. The ice is too weak to bear our weight and the water is so cold that we could not endure immersion in it. We would be trapped if fire should break out between us and the road to town. In the old days, our chances would have been negligible. Today we would have a grim wait for rescue by a Forest Service helicopter.

Every time the sky clouds over during dry periods, I listen for the sound of rain. Often it taps in a transient shower that does no more than strike rings in the dust. Sometimes it gushes in torrents that rush across the baked earth and drain into the lake, leaving only deceptive puddles that evaporate in the sun. Then, after a day when the milky lake has rolled steadily from the northeast, there is a pattering on the roof as of the feet of many mice. It quickens and steadies to a rumble—and the soaking rain has come.

I go out to feel the cool drops on my face. I watch the water darken and soften the dun-colored soil. Drying my wet hair, I look from the cabin window to see the branches bend and drip under the weight of the raindrops. Trickling moisture blackens the bark of the cedars and the pines. The

layer of fallen needles beneath them takes on a deeper
brown as it soaks up the life-giving water and lets it penetrate to the forest floor, and on down to the waiting roots.

In 1961, when our rain gauge showed only a half-inch of
precipitation during the six weeks preceding September,
many of the grasses and shallow-rooted plants dried up and
the tops of the big trees were drooping. Then Hurricane
Carla, passing far away on a northeasterly course, brought
eight inches of rain—five-and-a-half inches in ten hours. The
lake filled like a bathtub, while Ade and I cheerfully put
pots and pans under drips where the rain had driven through
our supposedly tight roof.

Three days later we had a second spring. Grass and buttercups and trilliums put up new shoots and, when the deciduous leaves began to turn yellow, dandelions showed out-
of-season gold on the ground. The mushroom crop was very
scanty, though, and the rains had come too late to help the
evergreens produce their cones. The chipmunks and mice
and squirrels would find the winter ahead a very hungry
one, for no artificial feeding gives them a proper diet.

Autumn begins here in September, heralded not by flaring banners of color, but by the "fall sog." The sky is overcast; the air is chilly and windless; foggy ghosts, coiling
downhill to drown themselves in the lake, eerily seem to pass
through unresisting tree boles; drizzle brings damp that
would make a frog rheumatic. Ade lights our first fire and its
smoke falls wearily to the ground, joining with the dampness
to add a smell of sodden burning to the dispiriting atmosphere. The exodus of summer people begins.

The wild things are in a fever of prewinter activity. I
dump the feathers of a defunct pillow into a box in the
woodshed and add a few newspapers for the convenience of
nest refurbishers.

A squirrel investigates, then pulls and scratches and yanks

mightily until she has detached a quarter-page of news-print. She grasps its edge in her mouth and tries to push it ahead of her, but it traps her hurrying hind feet. She and the paper do a forward somersault off the woodshed floor onto the ground. She picks herself up, chattering anxiously, and examines her tail, which seems to have been twisted in the fall. Tail all right, she turns again to the paper, now somewhat crumpled. A wild skirmish ensues, in which the paper seems to take on a life of its own. Gradually it is torn and wadded until the squirrel discovers that she can cram it into her mouth and, with caution, hop across the yard. At the foot of a black spruce she pats the paper into a tighter ball, squeaking and chirping the while, before she begins the seventy-foot climb up the rough bark. The paper escapes from her teeth when she is a third of the way up, and down they come, the paper to sail off into some maple brush and the squirrel to land flat and sprawled on the duff. She sits up, paws across her chest, and, stamping her hind feet, complains vigorously about such undeserved harassment. That over, she goes after the paper, wads it up again, and this time makes it to the bough that bears her nest. She looks fearfully down at me as I walk toward the tree to watch, but deciding, perhaps, that I am not a climbing creature, she carefully pats and crushes and pushes the paper into place.

Wondering how she knows that newspaper is fine insula-tion, I start for the cabin, but stop as a chipmunk, cheeks bulging with corn and mouth full of old-pillow feathers, skids to a stop in front of me. I stand perfectly still as she looks from one foot to the other and up, up, up to the top

of my head. Reassured, she jumps between the feet of the
colossus and scampers to the stone pile that protects her
burrow entrance.

The chippy reminds me that I must send out the order
for our own winter groceries before the freight stops its summer-only delivery. From the door, Ade is watching the
squirrel as she begins a second assault on the obstreperous
newspaper. "I guess it's time to seal up the roof," he says.
He gets a ladder, and I dig the grocery catalog out of my file.

Then a chilling wind clears the sky. The deciduous
leaves, their work done, spiral down. Frost turns the fern
fronds to buff and blackens summer flowers. Showers of
pale-brown pine needles trim the roof and weave intricate
carpets on the paths, hang like fringe on the fences around
the abandoned garden, and throw a network over the
dwindling squash plants. Pine cones thump on the roof and
fall to the ground. Cedar leaves the color of iron ore thicken
the layer of duff. When the wind's task is completed, the
trees are clear green again. The seeds of a bull thistle float
high on fall air whose tang creeps into the blood, to stir and
ripple and emerge in an effervescence of delight.

There are many wings over the water as the flocks of ducks
gather and the young ones practice take-offs and landings.
Often one, not quite strong enough, runs on the surface,
trying to lift with its brothers. Perhaps it will fly south with
them, perhaps not. If it cannot migrate, it will die. Nature
does not coddle the weak.

The ducks disappear, the geese fly over, and the green hills across the lake wear crests and brocades of gold. Ade and I, if our work allows leisure time, cross the huge rocks that guard our shore and take a boat ride. As we paddle out from the skid, we admire our mountain ash, its leaves a weaving of pale-yellow daggers against which its scarlet berries hang over the water. The gray jays are snatching the fruits, dropping many of them. The spilled berries bleed on the stones or disappear in little splashing fountains. There is movement under the water but the rippled surface is too deeply shadowed for us to see what lake dweller is feeding on such beautiful food.

The arc of the sun is shortening for winter and our south shore is in shadow. On the edge of the lake, maples flame scarlet and alders modestly display foliage of dusty rose. The smooth water is black-green, reflecting the autumn trees like searchlights turned down into the depths—a brightly colored inversion of the northern lights. Among the evergreens, the birches and aspens show pale green in sheltered spots, buttercup gold in the open, shining bronze where the big ones rise above young spruces, and rusty on the ridges where the leaves wait to fall in the wind.

The shadows of the past lie on the hills. Where glacial till alternates with ridges of rock, wide yellow swaths follow the soil from the water to the hilltops, and stunted cedars cling by twisted roots to crevices in the stone. Nearby stands a climax forest, grown through centuries and many cycles of grasses, shrubs, and various trees to full maturity. The stubby limbs of its black spruces, rising high above the aspens, are so thickly needled that the trees are top-heavy. Behind them, surviving white pines remember their fallen brothers. In the

distance a saw-toothed ripple bears witness to an ancient folding of the earth. The nearer hills, where evergreens and red deciduous treetops mark the slopes with hieroglyphic designs, are veiled ever so lightly with blue, while those more distant grow fainter row on row in a purple haze, the worn-down roots of the mountains.

We leave behind us the last of the lodges and cabins, and idle beneath granite bluffs, splotched with the maroon and green of lichens. In a little bay we come upon a beaver house. One beaver swims slowly across the water, its head leaving a rippled streak behind. We think we are unseen but, with a splash and a scattering of drops, its flat tail slaps the water in warning. We marvel at the swift, water-skimming flight of a pair of late golden-eyes. We paddle carefully through shallows where the jagged black teeth of the earth snarl up at us through the water. We stretch our legs on a secluded beach, where a stream pours out of a green tunnel.

The sun is low in the northwest. Ade starts the outboard and turns homeward over water that is flat and milky-blue in the evening calm. I watch the waves curling away from the boat in moving sculpture to collapse in a flurry of bubbles at the sides of our wake. At first the waves are blue with silver arabesques flowing over them, their bubbles clear and rainbow glinting, their scattered drops white like pearls or touched by shadow into spheres of blackness. Our trail upon the water lies behind us, gradually turning from turquoise to green. The slanting light is darkening from white to saffron to bronze, and details of the shore stand out in amber-traced clarity. The waves are green now, their patterns and droplets glowing like melted copper. The light begins to fade rapidly, and the waves are violet meshed with restless black.

In the west, a cloudbank is forming, rayed from its flaming edges by the hidden sun. A wind is rising and there is a chill in the air. Far ahead of us whitecaps flash across the ultramarine water like leaping fish. Ade opens the throttle and we race the squall home.

As we haul the boat to safety on its cedar-log skid, the wind reaches across the lake, beating the rolling water into foam-edged combers. Thunderheads are thrusting into a third of the sky and their voices rumble in the distance as the first lightning streaks over the forest. We run to the log cabin, ducking our heads against puffs of dusty, twig-laden air.

Unmindful of the weather, the tame squirrels are gathered in the yard for an evening snack after their long day of cone harvesting. We hand them graham crackers, while chickadees and woodpeckers, blue and gray jays, pick suet from feeders and snatch corn and crackers from the ground.

Suddenly the storm, cloaked in premature darkness, roars toward us through the trees and we and the wild things take cover. Inside, by the mellow light of oil lamps, we listen to the strange voices in the wind, to the crash and thud of waves breaking on the rocks. We hear the crack-crack-crack-snap of a big tree falling nearby. Ade blows out all the lamps except those near us, which we can extinguish quickly. If a tree should strike the house, we do not want fire to add horror to confusion. Torrents drum on the roof and pour from the eaves; beyond the black windows, the forest appears and reappears in lightning-white flares; thunder blasts and reverberates between hills and clouds.

Then it is over. The sky clears from the west. Water drops glimmer softly along the eaves as the twilight fades into night. It is time to light all the lamps, draw the curtains against the big dark, eat, and reopen the books we laid down the night before.

ch!

It is a good idea to look around before leaving the house

on these last warmish days when wild food is fast disappearing. Once I surprised a bear, contemplating the removal of a suet feeder, a nourishing bite before hibernation. He reared taller than a man; with his forelegs lifted and half-spread, and his body thickened by his heavy winter coat, he looked as broad as a door. I froze and he froze. After a moment's consideration we each turned around and went our separate ways.

In November, a steady, purposeful wind strips the last leaves and brings cold rain. The air is raw. Ears ache and hands chap. The chipmunks are tucked into their burrows until spring, and damp, mussed squirrels gnaw bits of heat-producing suet through the mesh of the feeders. The blue and gray jays spend more time around the house. The crows have gone, leaving a silence to be filled by the *crawnk-crawnk* of the ravens. While Ade and I are reading at night, enjoying the snugness of snapping logs and lamplight gleam on the handhewn walls, we hear a faint rattling against the north windows. In a beam of lantern light from the door we see shining rain drops, bouncing balls of sleet, and, here and there, a big, soggy flake of snow. We settle again in our big chairs. The house is tight; the fuel and food are stored. Let the winter come.

The advancing first snow hides the hills across the lake so that from our shore we look north into nothingness. Gradually the fine flakes engulf the house and climb the hill on the south, veiling everything in a semitransparency that is alive with the slanting movements of its particles. Here a flat stone, and the broad dead leaf of a plantain, turn white. Little tufts form in the crotches of the willow branches. A mass of tan reeds becomes an angular design in three dimensions as the white collects, here on the leaves, there on the ground beneath. The mass of the trees begins to take on light and pattern as each twig is thinly covered. In a few hours the dim world is changed to one of clarity. Tree trunks are sharply delineated by thick spatterings of snow on their north sides. The ground is inches deep in white, but still has the snowbent tops of tall weeds to break a smoothness not yet marked by tracks. The tips of balsam and spruce branches are turned into white-feathered ptarmigan feet.

Arctic air pushes south after the snowstorm and we wake to a temperature of around ten below zero, and to the short-lived miracle of hoarfrost on the land. The cedar leaves are edged with spiderweb lace; the branches and trunks, the weed tops and the garden-fence wire, are trimmed with fragile feathers; long frost fronds hang from the eaves, like the antennae of giant moths peering over the roof-edge. The clusters of green on the great pines by the cabin are balled with frost so that the trees are a giant's orchard in full bloom. The bare branches of a birch on a hilltop are covered with glassy needle shapes. The air has been very still to allow the formation of this ice, so frail that the breath of my passing shatters it, so fine that the heat from my skin melts it as I bend near.

With the day's temperature rise, the evanescent decorations vanish. The giant's fruit trees catch a rosy light, then slowly turn into pines again. The bare birch is hazed with vapor from the sublimation of its temporary silver hair, and faint rainbows flit among the twigs as the mist catches sunbeams through the branches. A slight and ceaseless shower of glittering frost, moisture freezing out of the air, falls like snow from the clear sky, and a shimmering layer of air waves lies above the lake as the great volume of water gives up its heat.

The freezing spray has glazed the shore rocks by December and, in the bays, there is windowpane ice that snaps and tinkles with the music of glass wind chimes. On a quiet evening we hear the improbable sound of barking seals. The ice along the shore, now an inch thick, has broken into plates, whose edges grind together and make the seal noises. The small floes pile up and the freezing goes on, until the lake is white and still.

This freeze-up pattern is one of our most regular seasonal effects but, as with all the happenings that depend on this land's capricious weather, the unexpected may wait just beyond the hills. In November, 1960, the thermometer dropped below zero and stayed there, day and night, while the wind pushed continually across the lake from the northwest. When the icy air reached our shore, it deposited as rime the moisture it had picked up from the rolling open water, and built up a panorama of splendor that would be a once-in-a-lifetime sight even for a native of the north.

At first there was a light tufting of crystals on branches and twigs, tree boles and rocks. The layer grew day by day until the twigs were inch-round and heavy with white, the tree boles wore vertical flutings, and the rocks lost their individual forms under their thickening cover. The wind still blew and the turbulent lake did not freeze over. Two weeks later, the tree trunks were frosted eight inches thick on their north sides; the living branches, their original intricacy lost under glittering padding, drooped like those of the weeping willow; dead and brittle limbs were broken off by the weight of myriad tiny crystals. Bushes collapsed into alabaster heaps, and the rocks along the shore vanished entirely in a bank of smooth and shining whiteness.

When the bitter wind finally died, I went out on a point of land and looked back along the shoreline. Except for the tops of the big trees in the background, the familiar forest had vanished. I saw only masses of white. Forty-foot balsam firs bent under balls of frost half as wide as the trees were tall, their branches and frost-coverings interlaced from tree to tree. Everywhere the shore was white, not the sparkling, unbroken white of fresh snow, but a whiteness of form and design, where pale flowers sprouted from ghostly earth and rippling, folded shapes were outlined by the palest shadows. For so pure was the whiteness that even the shadows seemed to be of a lesser whiteness than the unshadowed places. I was looking at a mass of living light, not bright with the touch of the sun and dark with the accent of shade, but made up of different degrees of whiteness, of different textures of light.

There was a tinkling as of little bells near my feet and I looked down at a thin sheet of ice along the edge of the water. The supercooled waters of the lake, at last released from their restless rolling before the wind, were starting to freeze.

As I watched, the ice sheet spread out from the shore, at first thin and rippling like transparent silk, then thickening and stiffening. Other sheets were reaching from the far shore and forming in the open water between. They spread and rushed toward each other, stirring the surface ripples in blue floods over their edges, there to freeze on top of the growing sheets. The movements of the great, flat planes, stretching like disconnected tiles over the lake surface, a half-mile wide where I stood and several miles from east to west, brought flashes of sunlight, as though from acre-sized mirrors. The lake, its surface as smooth as the stillest pond, gave an impression of heaving with the geometrical patterns in a giant's kaleidoscope.

And then the appearance of movement grew less. The blue areas of flooding water vanished. The mirror-flashes were gone. The sheets of ice had joined and sealed the miles of water for the winter to come. The whole marvelous change had taken place in little more than an hour.

Even as I watched, hoarfrost began to form in little buds on the clear, new ice. During the next three days the frost on the shore began to sublime into the now-dry air and the moisture was redeposited on the lake's ice sheet. The frost buds grew—into acres of crocus-like "flowers" and finally into frost "roses" as big as cabbages, their frail petals made of the most delicate and ephemeral crystals, so fragilely connected that the slightest zephyr would have destroyed them. In the below-zero cold, the warmth of the low sun did not affect them and they glittered amid their own blue shadows. Then came clouds and snow, and the frost roses were forever gone.

When the air temperature drops far below zero at night the ice thickens rapidly, especially if no heavy snows insulate it. The expansion of the solidifying water sets up enormous stresses that cause cracking, sometimes from shore to shore. Once when I was standing on the ice I heard a whining like the ricochet of a bullet. I felt a bump underneath me. While the whining fled away to end in a snap like the crack of a bullwhip, I looked down to see an inch-wide crevice between my feet, from which water was flooding around my boots.

When the ice is a foot or more thick, its cracking fills the air of bitter nights with sounds like the crashing of thunder and the slamming of big guns. Minute internal cracks and slippages join and lengthen, weakening the whole frozen layer. Eventually the ice breaks; there is a rumbling that grows louder as the crack approaches our shore. When the shock of the icequake is transmitted through the bedrock of the land, we feel a heavy jarring as the underground shock-wave lifts and lowers the cabin. The wave continues with lessening vigor until it dies out in the rock layer, its initial strength and the density of the rock determining the distance of its travel.

The ice on one side of such a break usually lifts above the original ice level, throwing up a pressure ridge that is some-times several miles long. The risen layer may be only a few inches high, producing a sort of curbing across the lake, or it may rise several feet, extending over the lower edge and leaving open water with broken chunks like little icebergs.

The blue-green of this ice puts to shame that anemic shade that fashion designers call "ice blue." Its blue has the bright-

ness of summer skies and its green has the richness of reflected moss and ferns. The very bubbles of the brooks and
the foam of rolling waves are caught within it, in chains and
clusters that the imagination can turn into the ghost-forms
of flowers and leaves. The underside is covered with glass-
like dendrites, faithfully reproducing the branching of trees.
"Blue ice" holds the shadow of summer past and the promise
of summer to come.

Such ice can support cars and even heavy trucks, but it
presents danger that is the greater because it is not obvious.
The pressure ridges are not readily seen when they are approached from the high side, especially when they are so
aligned that they cast little shadow. A snowmobile or snow-
sled could bump off a low one with slight or no damage, but
if it should run over the edge of a high ridge into the open or
thinly ice-skimmed water below, it would vanish into the
maw of the lake.

A "flowage," the local word for a series of small lakes connected by streams, has a strong current that flows throughout in one general direction, but that changes with channel
depth and width. The ice along the channel's sides is con-
stantly freezing and thawing under the influence of the
moving water. This varies the channel size and creates un-
stable currents that affect the surface ice. On one day the
flowage may be solidly frozen over. On the next, a current
may have undercut a section so that a breather hole has
formed, surrounded by a slush of flooded snow that is hidden
under a newly formed thin layer of top ice. Flowage ice is
never safe. Oldtimers test it carefully, pounding ahead with
poles that are strong enough to bear their weight and long
enough to cross the opening if they should break through.
Strangers may look on a flowage as a good trail for a snow-
sled and mistake breather holes for ice fishermen's locations,
often with chilly results.

Large lakes have under-ice currents, too, most often near

narrows or islands. They produce breather holes and thin

ice, the latter of which is the more dangerous because its thinness cannot be seen. Travel on ice, especially in motorized vehicles, which are heavy, and on snowshoes, which are clumsy, should be undertaken with caution.

By the first of the year the roads wait for the snowplow and air from the tundra brings deep winter. Now the whine of a powersaw three miles away may sound as though it were next door, and the dry air lets the morning's forty-below temperature rise to five above at noon. Although the shortest day is past, the sun is above the true horizon only from 7:30 to 4:30, and is so far to the south that it does not appear above the treetops. Only a few beams reach our windows and much of our light is reflected from the snow. The sunsets on the clear, cold afternoons are far to the southwest, but all the horizon is ringed with red that fades slowly to coppery rose and then into pearly dusk.

After a midwinter snowfall, the forest has the picture-card look familiar to everyone, but to be inside that picture is a far different sensation from looking at its printed representation. All around, in a windless silence, the snow spreads and flows and ripples, ermine-smooth under clouds and diamond-bright in the sun. There is a white-Christmas feeling that is unrelated to philosophies or carols, to merry-making or gifts: a feeling that endures the rumble of SAC jets and is undisturbed by threatening tomorrows. Maybe this is the "peace that passeth understanding," the happiness that men pursue so savagely among material things.

The moonlight, through the dry and dustless air, is so bright that even a crescent brings shadows to life. I stood on the lake ice one afternoon, bemused by the sight of gray sun shadows slanting from the west, crossed by purple full-moon shadows from the east. After the twilight fades, the moonlight turns the icicles on the eaves to translucent blue fringes and strikes sparks from the frost platelets on the snow. The clearing is mystical and strange until I see movement at the edge of a patch of shadow, and make out our tame doe, patiently waiting for some protecting darkness to cover her feeding place.

Unless the sky is heavily clouded there is always some light in the winter night—the bluish aura of the moon, the low, green glow of the aurora, the scattered pinpoints of the stars. They blaze against the clear, black sky and sometimes shine through the mistiness of the aurora like jewels on the veil of Time. The red and blue giants gleam pink and azure amid the twinkling whiteness of their companions, and the planets carry steadily glowing lanterns along their ancient paths.

The aurora shows a dim light low in the northern sky all year, but reaches its splendor in winter when dry air and long hours of darkness make way for its fabulous light. Sometimes it flames red as the hearth fires of Valhalla or green as the waves that bore the Vikings' ships. Or its cold-white banners and curtains strengthen into searchlights that set the birds twittering. Once I saw a squirrel and two chickadees come looking for breakfast by aurora-light, only to depart in confusion when the "dawn" faded from the sky.

There was a night when I woke to windows filled with a pulsing, unearthly green. Hastily wrapping myself in clothes heavy enough to defeat the below-zero air, I went out into a night that one might be more likely to encounter above the Arctic Circle. The wild green flare lighted my way as I skittered down the hard-packed snow of the path onto the wind-ruffled drifts that covered the lake ice.

A mile to the north, the Canadian hills crouched low before a cold and flaming glory. Two silver arcs towered above the rolling horizon like a faded double rainbow, a shimmering band of platinum between, a crystal-flecked velvet darkness beneath. Rays of white and rose swayed high above the arcs and the wavering green overhead seemed to come from everywhere. And, in the quiet night, my ears caught the faint swish and rustle, like lightly touching taffeta ribbons, that is the voice of the aurora.

A distant hum came from the west and grew into the throbbing roar of an airliner, approaching on an emergency route. It passed directly overhead, flying low, sleek as a fish, gay with its red and green and white lights. I wondered whether anyone beside its crew was watching the north with me. The plane flew out of sight, an incongruous reminder of a civilization so much more remote than mere distance could make it.

The lights in the sky grew dim. The platinum slowly lost its sheen and the lower darkness grew larger and deeper and seemed to come terrifyingly nearer—a black whirlpool, utterly without light and of endless depth. As my eyes tried to penetrate this opening into space, the northern lights faded. Only a faint glow lingered to outline the hills, and the gray distances stretched on and on.

The Trees

ONLY ONCE have I seen a big tree reach its natural end. It was the largest tree around here, a spruce from the old days, four feet thick and almost 140 feet tall, that stood some twenty feet from the summer-cabin porch. Ade and I were lunching out there on the day when a squall roared in and split the spruce a dozen feet up from its roots, so that with every gust of wind, the break opened and closed like a big spring. The tree leaned over the bedroom; we would get someone to cut it down the next day. Taking out a tree that size and so close to a building was no job for an amateur.

In the blackness before dawn, Ade shook me awake. There was a howling of wind—and the heavy, steady cracking of the spruce. We grabbed our robes and, barefoot and shaking, tumbled through the door and ran to a safe distance. Ade threw a beam of lantern light across the roof.

The gale from the west was so strong that my sleep-addled brain could not quite grasp why the beam of light did not bend before it. Everything else did. The sky was clear and bright with stars, and the aurora glowed faintly.

The snapping, cracking, yielding of the spruce's fibers sounded through the noise of the wind in the branches. The tree bent, swayed, leaned until its top was more than halfway down, but still it sprang back. Its fight was terrible and sad. The first faint rays of dawn were on its top when the lonely struggle ended. During a pause when the wind was gathering strength, the tree began to sway. Slowly, with the dignity of the gallant defeated, it tilted. Its battered trunk parted with a great splintering and the tree fell, tearing branches from its neighbors before it stretched out on the earth. And the whole forest stood silent and watchful.

It fell away from the house without destroying a single one of its smaller, close-grown neighbors, as though it had been guided. And who is to say that it was not?

When a tree like that is gone, there is nothing you can do about it unless you plan to stay another couple of centuries. I counted 187 rings on a cross-section of its trunk, the two outer inches of which had been destroyed by insect borings and fungus. The big pines, some of them three and a half feet in diameter, are as old. The Forest Service measures the thickness of a tree at breast height, about four and a half feet from the ground, and the average diameter of an eighty-year-old white pine is only thirteen inches. When Ade and I had the opportunity of receiving electric power and telephone service at the cost of felling a swath through our old trees, we decided in favor of the trees.

The economy of the first white settlers in northern Minnesota was based on the pines that only seventy years ago marched in phalanxes across the hills. The areas that escaped devastation by lumbering and fire are few and small. Some of these are under Federal protection within designated "primitive areas." Others have been spared by private owners, or are so isolated that lumbering was too costly or difficult.

East of our home, on the south shore of the lake, is one of these remnants of the past, where white pines cover a ridge and stand in a majestic row along its top. Their branches, as large around as many second-growth trees, reach straight out from the trunks to touch their neighbors' needles, which, on this species of pine only, grow in clusters of five, one for each letter of w-h-i-t-e. These trees lean toward the southeast, away from the prevailing winds, and the branches on their southern sides are longer and fuller than those that thrust into the bitter northerly gales. One by one, they are going. Erosion-weakened roots let a giant lie down to rest. Lightning, striking into a hollow trunk, destroys another. Savage gusts set a top whipping on its tall stalk, to snap off and leave the stub to disintegrate through many years. One summer, we smelled smoke and found one of these great pines burning at the base. Campers, who had built their fire in the hollow between two of the roots, had not bothered to quench the embers. The fire was put out, but the tree was killed.

Two white pines, at the edge of our log cabin's clearing, are two and a half and three feet thick and more than a hundred feet tall. The giants of the old days were sometimes twice these dimensions. When I am out of sorts from niggling little things (which exist here as in any other place), I stand on the roots of one of these trees, leaning back against its rough bark. The trunk rises, straight and true. It seems too solid to move in the wind, but I have seen it sway. Fifty feet up, the branches begin their spread and make a pleasant, latticed roof that does not hide the sky, where, on cloudy days or bright, gulls sail on joyful wings. The biscuits that turned out to be hardtack because I forgot the baking powder seem quite trivial.

Deep in the untouched part of our land stands a handsome representative of the Minnesota state tree, the red pine, which is also called the Norway although it is native only to North America. This great tree's bark shows irregular diamond shapes of rusty red through its gray cork. Its columnar bole is topped by a thick crown that looks as though its branches might be clothed in dark-green fur. Its fallen, paired needles are eight inches long.

Ade and I make special boat trips to see small red pines on the north shore of the lake where they have ample sun, space, and water, for they are as beautiful as anything that grows here. Their colorful boles reach up only a dozen feet or so, spreading close-set, sturdy branches. Every twig is thickened to bolster size by the masses of long needles. In the spring, their new growth rises like pale candles; in the fall, their tops are alight with cones; and, on every summer day, the aromatic rosin of their branches perfumes the air.

The pines outlast the towering black spruces that look immortal but, with their short, thickly needled branches, offer much wind resistance. Ade and I, on a day of scudding clouds and sporadic breezes, decided to transplant sweet williams from the paths, where they seem to thrive best, to the south wall of the log cabin. We had started to dig when the driver of the gas truck came to tell us that one of our spruces had fallen across the road.

He had an ax. Ade got his chainsaw. I hauled out cut-off sections. Within a half-hour, the road was clear.

Ade and I started toward the cabin. Another spruce lay across the path. A third was tottering fifty feet ahead. A hundred feet above the deceptive ground breeze, a heavy gale, almost soundless because it topped most of the trees, was catching the exposed tops of the spruces and snapping their boles like twigs. This was no day to be blithely digging up flowers. Ade tells me that I gave a fine imitation of a startled hare as I dashed toward the relative safety of log walls with both hands over my head, presumably to protect myself from trees that weighed several tons. Eight of them fell in our yard on that afternoon.

The pointed tops of the white spruces and balsam firs give a year-round Christmas-tree effect to the forest and much of the fresh, woodsy odor comes from balsam rosin, which exudes in blisters on the bark. It is sticky—harder to

remove from the skin and clothes than tar—but it smells so good that one is inclined to be tolerant. Ade felt differently, though, the time he leaned against a tree and was caught by his hair.

Balsam needles grow in flat rows along the sides of the twigs, while spruce needles swirl around the twigs. This gives balsam seedlings a sparse, angular appearance, while little spruces look furry. In spite of their soft appearance, spruce boughs make prickly camp beds because of their up-right needles, and they are short-lived as indoor Christmas trees because their needles fall rapidly as they dry. Balsam does better on both counts.

Because balsam and spruce trees grow a new row of branches each year, the age of small ones may be told by counting the whorls. The new balsam tip sprouts straight up from the point that was the previous year's treetop, with the new branch whorl spreading from the same point. The spruce puts out new growth like tentacles, also from its former top, that writhe and twist until one of them takes precedence and grows vertically as a top extension. Then the others settle down and form the whorl of branches.

Only the first sprout of the northern white cedar, which looks much like the sprout of fir or spruce, has needles. The second sprig shows the lovely spray of overlapping green leaf scales. These trees grow very slowly and take many different shapes. In the open they may keep a ground-to-tip conical form, but this does not happen here because the deer nip off their lower branches. In the log-cabin clearing, where there is not much water, the cedars, although more than a century old, are less than a foot in diameter and, with their bare lower trunks, are shaped like sugar maples. By the brook, where the ground is swampy, the cedar trunks are thick and covered with spiky branch stubs, and, by the summer house, there is a very tall one with only a tuft of green at the top. Years ago a bear climbed this tree, lost his grip, and came thumping down, bringing most of the branches with him.

We have many cedars around the log house. Their low limbs are sheltering and friendly, although they bend so under the weight of snow that one must duck or get a face-wash. The boles of a pair of them are curved gracefully away from each other like the frame of a lyre; they sprouted and grew from under a glacier-dropped boulder that was rolled away by someone years ago.

The cedars' heartwood may rot; erosion may undermine and tip them; storms may twist and polish those on the lake shore into bleached grotesqueries; but somewhere a root still takes sustenance from the earth and somewhere leaves stay green on the battered branches. They are living symbols of tenacity and endurance.

Squirrels strip their red inner bark for nest linings and store their little cones, which are full of nourishing seeds. The bronzing of some of their leaves reminds us that it is time to start the fall chores. And the cedars are beautiful with their softly ridged gray bark, their yellow masses of half-formed cones, their twisted roots with mysterious hollows under them where wild things find homes and shelter. Whenever someone tells me that he has cut thirty cedars before he found three that were sound enough for posts, I wonder that he did not bore them first to learn if they were hollow. When someone cuts these trees indiscriminately to feed the deer, I wonder that he did not lop off branches here and there instead, that there might be deer food in the future. Perhaps the pines, with their impressive size and tragic history of wasteful cutting, take more than their share of admiration. One has missed much who has not known a white cedar tree.

Near our home is burned-over land covered by tall, slim, second-growth timber, much of which is paper birch. It is also called silver birch, from the sheen of its bark, and canoe birch, because the Indians used the bark for canoes and other waterproof vessels. In its early years the trunk is a rich, dark brown, with many short, pale, horizontal markings, and looks much like that of a cherry tree. When a young birch grows past the sapling size, its outer bark-layer splits and peels away to show the familiar silver-white, many-layered coat, with its large lenticels that legend says are pictures of the thunderbird, who brings storms to the northland.

Nanabojou, a great Indian magician, once hid in a hollow birch to escape the thunderbird, whom he had angered. Since the birch is one of the thunderbird's special children and the storm-bringer could not send his lightning bolts against the tree without injuring it, Nanabojou escaped. To thank the birch for saving his life, he put pictures of the thunderbird all over its bark so that it could be easily recognized and would never be struck by lightning in error. (This magic seems to have deserted the birches in our area, for two of them less than a hundred feet from the cabin were struck last summer—within five minutes—at the height of a violent thunderstorm.)

There are not many birches among our virgin evergreens, but the old ones that grow there are as tall and as old as the pines. They brandish their white, dead, upper branches against the sky. They wear their stained and tattered bark with dignity. They ignore the fungus that powders their wood and shelves their trunks with brackets. Even as the wind brings down their defiant upper limbs, new leaves sprout on those below and catkins and buds flourish as in the trees' prime.

In 1954, Ade considered for firewood a huge birch that was dropping dead branches on one of our paths, but let it stand when he saw no way to fell it without destroying a strong young tree that had sprung from the old one's roots. Now the big tree leans so sharply that I have to duck when I walk under it. It and its offshoot have twisted progressively so that the shoot is almost clear of the parent tree's line of fall. When the old tree smashes down, we will see whether nature has spared its offspring.

When we first moved here I was puzzled by "popple," a colloquial name for a tree whose species seemed somewhat of a mystery even to many long-term residents. I have since learned that, in this area at least, "popple" usually means quaking aspen, whose almost-round leaves, quivering on flat stems in the slightest breeze, are a quick means of identification. The aspen sapling's bark is smooth and gray. As the tree matures, the bark is pale, sometimes almost white, and so marked with lenticels that it can be mistaken for the birch. In overmaturity, the trunk bark darkens and ridges from the base upward until it resembles that of the white pine, but the branches are still light. This aspen, usually described in the literature as smallish, may, in old woods, grow as large as the birch.

"Popple" also means the balsam poplar, whose leaves are less round and quivery and whose showers of white-haired seeds lead to its being called cottonwood, a name usually applied to a tree that grows along watercourses in the plains. As if this were not confusing enough, someone recently told me that "popple" was "bammagilly." After taking a deep breath, I realized that this meant balm-of-Gilead, originally the name of an African or Asian evergreen, but applied here to a sterile poplar that reproduces by shoots. Again, this tree is sometimes classed as a variety of balsam poplar, and it is also called balm-of-gills. I think that, somewhere between the African balm-of-Gilead and the "bammagilly," the balsam fir, whose rosin has medicinal properties, influenced the picture. If I had it to do over again, I would let the whole thing go as "popple."

It is easy enough to see how balm-of-Gilead became "bammagilly," but "popple" is more than a simple corruption of poplar. In the old logging days, any tree whose wood was weak and snapped easily was called "popple" because it popped. These weak trees were mostly of the poplar family. The lumberjacks avoided using such wood in construction designed to take strain, because it was dangerous. A superstition developed, and many jacks would not shelter in a building that contained "popple" logs. There is some sense in this, because "popple" logs and lumber are susceptible to insect and fungus attack and do not hold up well under damp conditions.

Near the summer house, small aspens and pin, or fire, cherry seedlings form almost a thicket. The cherries have all seeded from one tall, slender tree. When its top is a cloud of white, interspersed with the few leaves that opened first on the dark twigs, I can easily imagine that the petals are the dancing skirts of fairies, washed in dew and hung out to dry, as an old Irish lady told me quite a few years ago. Later, when the sour, red fruits are ripening, the trees will be aflutter with birds.

Around a little cabin near us, a gentle old man named Charlie Olson chopped down the central trunks of bushy chokecherry trees to pick the fruit he could not reach. This increased the number of shoots from the roots and the trees, now shrublike, bend under long clusters of cherries. In the same yard, a legacy of old Charlie's appreciation of beauty, is a neatly trimmed, symmetrical mountain ash. Its flat bunches of white flowers lie like bouquets among its dark-green compound leaves and the massed red berries that follow are as lovely against the pale buff and rose of the fall foliage. Many seedlings, so lacy that they look like ferns, sprout on the forest floor from seeds that have passed unchanged through birds' digestive tracts, but few of the

43

seedlings get enough light to be successful. Those along the sunny edges make a noble effort each summer, but the deer are fond of the twigs. One can get as much satisfaction from watching a fawn nibble the young growth as from looking at the flowers and fruit.

A formal gardener would clear the brush that dots our land, and local residents usually remove bushes to increase ventilation, but we cut only a minimum because brush is very important in the ecology of the forest. Deer, moose, and hares browse on its leaves, buds, and twigs. Birds nest in it and feed on its seeds and fruit. It offers cover to all wild things, from the smallest shrew, scurrying under a carpet of its fallen leaves, to the biggest bear, bumbling along like a fat shadow between its stems.

brush" is important

Scattered in open spots are the look-alike striped maple, or moosewood, and mountain maple. Both grow in bushy form and have shallowly notched leaves, but the greenish bark of the former is striped vertically with white and that of the latter is not. Mountain maple and red osier dogwood, which also goes by the intriguing name of kinnikinnik, are favorite winter deer foods. Studies made by the Minnesota Department of Conservation give both of them high ratings for nutritious elements.

On the hill south of our log cabin is a flourishing growth of red maple. Its light green or gold and red are pleasing seasonal contrasts to the dark conifers. This maple has smooth gray bark, velvety red twigs and buds, leaves that are whitish beneath, and creamy flower clusters. I wondered about this fifteen-foot, brushy form of a plant that usually grows into a tall tree, until I noticed that the shoots grew more or less in rings from the edges of the decayed remnants of broad stumps, cut so long ago that they have almost vanished into the earth. The big red maples were the victims of the early settlers' desire for hot-burning firewood.

Keys from the maples start sprouts everywhere that sunlight falls. They form a scattered undergrowth beneath the big trees and, if we did not "weed" the clearings, we should soon have to cut our way through. This new growth is good, for the deer grow fat on the twigs.

The gentle whitetails move slowly through the woods and across the hill, unhurried and, to our joy, unafraid. In summer they browse on leaves. Later a snap signals their presence as a doe breaks a tall shoot down to within the reach of her newly weaned fawns. They tear the twigs off in a leisurely manner, leaving twisted, ragged ends because deer have a hard, horny pad in the front of the upper jaw instead of teeth. When they have filled their first two stomachs, they relax. A ball of food slides visibly up the doe's throat and she chews her cud, preparing the food for passage to her third and fourth stomachs, where it will be digested. One of the fawns lifts a front leg and bends it back, balances carefully, and lies down. Soon all three are lying quietly, chewing and resting, their neutral coloring blending so perfectly with the forest duff or the snowy brush that they are almost invisible unless they move.

Another shrub that is browsed somewhat by deer and, in summer, by moose is the speckled alder, which goes by the common name of tag alder. The word "tag," which designates any small pendant thing, has come to mean a catkin. The alder, like the related birch, has two types of catkins. The green hanging form is flexible and made up of many tightly fitted, exquisitely fragrant flowers. Their pollen spreads a skim like yellow snow on the spring earth and fills the water with golden glints. The other catkin is a small green cone that turns brown and opens when its seeds are ripe. The alders like water. They grow among the maples on our hillside near the outlet of a spring, along the lake shore, and beside streams, where they hold the earth against erosion.

[handwritten margin notes: deer have 4 stomachs; chew cud; 3rd & 4th where digestion takes place]

Next to one of our shoreline alders is a tall Juneberry that is covered in spring with drooping white flowers, their five petals flaring and widely separated. It bears edible purple-black fruits that I have never tasted because the birds always get there first. This bush is also known as the shadbush or serviceberry, and is a member of the rose family. I will not hazard a closer identification because even botanists are uncertain of the identifying marks of some of the Juneberries.

Two of the many willow species, which are also hard to identify, grow outside the log cabin. The glistening, furred pussies of one of them are the first-opening buds of spring. These pussies become white flower-clusters and other buds on the tree put forth tentative metallic-white hairs before their clear-green leaves emerge. The second willow presents its pussies later, but makes up for its slow show of life by bearing clusters of bright-yellow blooms. One fall a buck chose its small stalk as a convenient post for rubbing the last of the velvet from his antlers. Six months later I looked at the shredded bark and thought the tree would surely die, but the little willow was strong. In spite of its almost girdled trunk its top glimmer of buds gave place to a luxuriant growth of slender, glaucous leaves. One should never be hasty in cutting a tree.

We do not agree with some of our neighbors who consider the red-berried elder a weed. True, without its leaves, it is twisted, irregular, and sprawling, but it leafs out full and thick and grows so rapidly that a small shoot in a favorable location will be a ten-foot bush in five years. Its flowers are like white lilacs with inconspicuous petals, and its bunches of berries hang from almost every branch tip. We have two types of this elder—one with green twigs and holly-red berries, the other with purplish twigs and dark-scarlet fruits. There is a big, tangled, bare elder outside of the window where I am writing. Its fat purplish buds are just starting to swell and a buck is running his lower teeth up the stem, looking very pleased as he scoops up the buds. The elder will grow lovely in spite of the deer's feeding and I will watch thrushes and grouse stuffing themselves with the berries.

Many other types of trees and shrubs are found in the Superior National Forest and in private holdings, like ours, that border it. This national forest covers 4,100 square miles. It contains a "no-cut area" of 362,000 acres along the Canadian border within a "roadless area" of 1,038,700 acres.

These acreages are not waste lands that no one ever sees and that do no good. The "primitive" or "no-cut" area is held in its natural state as a sanctuary for those who seek the peace of wild places and as a control laboratory and guide for foresters and conservationists. It is never lumbered except for possible salvage purposes. The balance of the "roadless area" is subject to lumbering, but no cutting is permitted within four-hundred feet of shorelines, trails, and portages along the canoe routes followed by vacationers, and only temporary roads to logging operations are permitted. The traveler on the public roads that border this area sees only miles of unbroken forest and may get a false impression of vast, unharvestable timber reserves. Other commercial operations are banned to prevent the destruction of this great canoe country.

Our home is at the edge of a section of 593,000 acres of the Superior National Forest contained in Cook County, Minnesota. The terrain is hilly, with rocky outcroppings. The soil is glacial till, coarse and with little organic matter. These features, in combination with a growing season of only a hundred days, make this inland forest unfit for agriculture. There is sand, gravel, and building stone, but the last is mostly dark granite not much in demand, and the location is too remote to make quarrying and shipping economical. There are mineral deposits, but those found so far are not, as rumor has it, huge accumulations like the iron ores of the Mesabi range. They are so small or of such low grade that mining would be as economically unsound as quarrying. The main economic value of Cook County's inland area, and other similar preserves, lies in the timber and recreation potentials. Toward the development of these the Superior National Forest's "roadless areas" are being managed, and managed very well.

In December, 1951, these "roadless areas" were closed to all airplanes except those in government or emergency service. This brought complaints that the ruling prevents the enfeebled, the very old, the very young, and the disabled from enjoying wilderness. The argument is meretricious because, although it claims that a special group of people are being denied the benefits of wilderness, any place within the "roadless areas" that might be developed to permit the landing of planes and the entertainment of guests would no longer be wilderness, which is a place where man comes and goes, leaving no permanent developments. Any new establishments would only duplicate the accommodations on the outer edge of the "roadless areas." Throughout the country, special-interest groups who would like to exploit the irreplaceable, publicly owned wilderness lands for short-term, and often private, gain, use such propaganda, which appeals to the listener's emotions and ignores his intelligence. Fortunately, no one has yet recommended that the Rocky Mountains should be leveled in order that those who cannot stand high altitudes may ride safely over the rubble heaps.

Aside from the danger of a crack-up in remote forest and the difficulty of finding the wreckage and succoring any survivors, planes add to the danger of forest fire. The sunny summer days, so much desired by resorters and vacationists, are no joy to foresters if they continue very long, for each added day without rain increases the fire hazard that increases still further in proportion to the number of campers and others in the forest.

The holocausts that swept through the Canadian forests during the drought of 1961 are a matter of hideous history, but no account can possibly picture the horror of forest fire—the timber losses, the towns cowering in smoke before evacuation, the desperate struggles of the fire fighters, the herds of deer trapped and roasting alive, the small things dying awfully in boiling pools, the whole ghastly destruction of life and the potentials of life for years to come.

During the time of the great Canadian fires, our sky was dimmed by a yellow smoke haze that blurred the hills and hid the ends of the lake in murky shadow. A red sun rode across the sky, and we could smell the smoke. I heard shouts and chopping as men of the Forest Service put out a small blaze only a mile away. I watched smoke rise from the Canadian hills, followed by white puffs of steam when water, scooped by plane from a small nearby lake, doused the flames. It is fortunate that the day was calm, because a wind can carry even a little fire out of control in a very short time.

Although many forest fires are caused by lightning, the two within sound and sight of our cabin started from camp fires that had been left to "burn out." Of course they will burn out—and take the forest with them! An evergreen forest, even after six weeks without rain, still looks green and fresh. But, if you pick a dandelion leaf, it crumbles in your hand instead of feeling smooth and pliable. At such a time, every spark may be a source of disaster.

Ade and I do not smoke outside in dry weather. Even when it is damp, we never put out cigarettes in the duff but grind them to pieces on a rock. When we build a fire outdoors, we look for bare ground or rocks. If we cannot find such a place or clear one, we soak the earth with buckets of water before we build the fire that we keep small so that sparks will not fly into the woods. We do not throw a bucket of water on it and go away. We dig into the wet ashes, and often find smoldering embers. We dig and soak until everything is wet and cold. And we always heed the warnings of that famous Forest Service spokesman, Smoky Bear—"Break your match! Drown your fire! Everybody loses when timber burns!"

The Fungus and Its Partners

I STARTED from the summer house to the log cabin on an
afternoon when the sky was hummocked with soggy-looking
clouds, one of which dumped itself in streams as I came to
the edge of the woods between the houses. I ducked into the
open woodshed and sat on a log, looking through the rain.

Against the trunk of a big birch the peeling bark lay in
vertical rolls, like parchments whose markings held the history of the forest. The other trunks looked mottled in the
dimness—the striated gray of the cedars, the blistered green-
gray of the balsam firs, the ridged brown of the white pines,
the scaly gray of the spruces. When the shower passed, the
lichens that patterned the bark stood out against a water-
soaked black background.

Flat against a cedar's trunk were greenish pancakes with
crinkled edges, and irregular gray patches with tiny, deeply
lobed extensions. A balsam branch was covered by a straw-
colored, leathery complexity, its uplifted branches less than
a half-inch tall and intricate as coral. From the branches of
the firs and pines hung gray masses like tangled hair. The
stump of a fallen spruce was swathed in flat and nubby
gray-green, from which cone-shaped fairy goblets lifted; some
of these cups had perforated patterns and lacy edges, as

though they might have been made of twisted snowflakes mounted on minute stems.

On the edge of the woodshed roof, miniature scarlet pillows rose on upright stems from a gray growth that resembled felt. On a boulder beside the path, a flat black sheet clung, like charred paper whose curled edges showed a pale underside. Beside it was another flat growth, dark green above and made up of several loosely waved "leaves" that were dull black underneath. On the surface of a half-buried rock ledge, splotches of dusty rose were almost concealed by a spreading white fan—all looking as though they had been painted there.

The identification of the hundreds of species of lichens is a difficult business, requiring special knowledge, and often microscopic measurements. Thus, common names are applied to groups of species that have the same outstanding characteristics. The numerous lichens with scarlet tops are "British soldiers"; those with cups are "pixie cups"; gray-green, crinkled flat growths on trees and rocks are "curly crust." A red-brown form that colors the cliffs along the lake is the "cinnabar" or "flame," although the latter is usually a nickname for a common, bright-orange lichen. The gray, upright types that cover sandy, northern barrens are "reindeer moss," and the hanging tangles in this woods are "old man's beard," or "deer moss" because the whitetails eat them. "Deer moss" looks much like the Spanish moss which hangs from trees in the south. "Spanish moss," however, is neither a lichen nor a moss. It is an epiphyte, a plant that takes all its nourishment from the air. A flat black lichen that grows on stones is "rock tripe" and, I am told, can be boiled and eaten. Since I do not know which of the similar forms is edible, and since some lichens may be poisonous, I am happy to forego a lunch that looks, and probably would chew, like rubber.

Every lichen is composed of two plants, a fungus and a

green alga, working together in a mutually beneficial association. The fungus anchors the partners to bark, wood, rock, soil, or other substratum, and grows around the thin-walled, one-celled alga, protecting it from dehydration and death. The alga, by means of its chlorophyll, makes food for itself and for the fungus from air, water, and the energy of the sunlight that filters through the fungus body. Upright stalks, more correctly called stipes, bear the spore-producing organs of the fungus half of certain lichens. Spores are also produced by small "disks" that rise slightly above the surface of the prostrate forms. Such spores are worthless, as these fungi have become so adapted to their partnership that they cannot exist apart from their supporting algae, and the spores, of course, cannot reproduce the algae. Lichens are spread by windblown particles, made up of a bit of fungus body, surrounding a few algal cells. Perhaps the spores are carry-overs from primordial times before the fungi entered into successful, and dependent, partnerships with the algae. On this partnership and its individual components rests the whole economy of plant and animal life.

One-celled algae, along with bacteria, are believed to have been the first life forms to appear in the waters of very ancient times. Remains of one-celled plants have been found in the Gunflint formation that slants northeastward three miles west of our home, where rocks have been dated by means of radiocarbon tests at approximately 1.7 billion years. Eventually the individual floating algae combined into chains and attached themselves to the bottoms of shallow streams and ponds. I look with respect at the green "scum" on stagnant water and at slimy "water weeds" waving in a stream, for not only have they survived with little change through all the climatic and geologic modifications since the days when the earth was surfaced only with water and barren igneous rock, but their forebears gave rise to all the plant forms that have grown or still grow on the earth. If higher life as we know it were suddenly to vanish, these humble algae might still live on to start a new life cycle.

The fungi, which have no chlorophyll and must depend on outside sources for food, probably arose in those ancient waters, and the first alga-fungus combine may have developed there. Or a bit of fungus, or a spore, and some algae may have been washed onto a rock and, after many trials, clung there in the form of the first lichen. Acid by-products of the lichen's life processes attacked the rock and it began to decay. Bits of the lichen broke off and caught in the decaying stone, there to rot and add to the formation of soil, which would someday be ready to sustain great forests and to produce our food. Today, in the heat of deserts and the cold of the polar regions, on the poorest earth and the barest rocks, wherever the air is pure, the lichens nibble away, slowly, patiently—creating the nuclei of soil for the forests of the future.

That fungi can grow on almost anything (some even attack glass and metal) was demonstrated to Ade and me when we came to spend our first vacation in the log cabin. We opened the door to an atmosphere so spore-clouded that it set us sneezing. A breadbox revealed a flourishing green fuzz on a forgotten crust. Patterns of black covered the bare log walls and yellow spots blotched the painted kitchen counter. Dead flies were stuck to windowpanes, surrounded by haloes of spores released by the fungus that had digested their interiors. Mushrooms grew upside-down on Celotex ceiling panels, kept watered by a roof leak. After days of cleaning, airing, and drying (by means of fires that turned the North Woods coolness into an oven), we had a livable atmosphere, except near one door where the musty scent lingered. In desperation, Ade stripped off the frame and found several pieces of a shirt, stuffed away for heaven-knows-what reason, slowly being consumed by a pink-and-black garden.

The fungus body, or mycelium, is composed of a complex growth of threadlike tubes, so fine they are not individually visible to the naked eye. These tubes penetrate the material from which the fungus draws its nourishment and often are entirely hidden. The fuzz that we call mold is made up of the many fruiting bodies of a fungus, from which enormous numbers of almost invisibly small spores are discharged, each with the potential of starting a new plant, provided it reaches a suitable substratum.

The tough brackets that grow on the trunks of older birches in the forest do not kill the trees. It takes several decades for the fungus in the tree's wood to produce its first bracket. The tree may live on for a century, its wood gradually becoming powdery as the fungus feeds, and its trunk bearing more brackets with the passage of time. This fungus is of the family of Polypores and the many small pores from which the spores are discharged are easy to see on the underside of each bracket. In the days of flint and steel these brackets were boiled, dried, and scraped to produce excellent tinder.

Other fungi enter trees through the roots or through wounds like frost cracks and trail blazes, and hollow out the inert pith, or heartwood. I have seen ovoid white spots of thickly woven mycelium in the spongy heartwood of old aspens, windthrown and cut up for firewood. The "red rot" found in white pines produces similar moldy spots in pith that it turns reddish. Some of our white cedars whose centers are completely hollow grow as though undamaged.

Ade and I are always on the lookout for "dry rot," which is really caused by sealed-in moisture that permits fungus to reduce a piece of wood to dust. It is pretty startling to put your foot through a rotted floor board—even more so if the board is in the bottom of a boat. We are fortunate that we moved into our cabin in time to prevent the multitude of fungi from invading the logs.

We also watch for the dreaded blister rust, a fungus that attacks the needles of the white pine and develops a canker that works slowly down the twig and the branch, finally girdles the trunk and kills the tree.

This disease does not spread from pine to pine, and an affected tree need not be destroyed. The tree may often be saved if the canker, bleeding sap and rosin, is discovered on a branch and the branch sawed off while the damage is still five or six inches from the trunk. Nor is the fungus' attack limited to young trees with tender bark. A survey made under the auspices of the United States Department of Agriculture showed that during 1948 through 1952, in Minnesota, Wisconsin, and Michigan, blister rust infected trees from one to twelve inches in diameter at approximately the same rate.

This rust begins its life on gooseberry and currant bushes, both wild and cultivated. From them, the spores are blown to the pines, sometimes as far as nine hundred feet. Ade and I pull out the few plants of this type that we see in our yard, for blister rust cannot develop in their absence. A single bush that would produce only a handful of fruits for the birds could be the means of destroying pines the like of which may never grow here again.

Like the brackets on the birches, mushrooms are fruiting bodies, extending from hidden mycelia. In wet weather they pop up from soil, wood, or duff in a bewildering variety of sizes, colors, and shapes. On a dead branch are fluted awnings, cream-colored, with bands of orange trimming their edges. Poking up through a thick layer of pine spills are hundreds of three-inch stems that look as though they would surely open into something else, but that merely widen slightly at the top. Nearby is a fairy ring of dun-colored coral, with forked, hard branches. All alone is a ten-inch umbrella, covered with raised reddish spots. Here is a clump of little puffballs, some round and white, others burst in the center and free of their spores. There are gray eggs on stems, brown parasols of a size for mice to carry, convoluted growths that look like brains. One night I saw a faint blue light near a stump and found a clump of glowing mushrooms that gave off faint heat to my hand. (This glow is a form of bioluminescence, which has been observed in fungi, bacteria, and various lower forms of animal life. It is produced by chemical reactions that are best understood in the instance of the firefly.) These mushrooms, not edible, are bright orange, and the combination of color and ghostly emanation has given rise to their common name, jack-o'-lantern.

Among all these are edible species, but I will not attempt to identify any of them. Eating wild mushrooms can be extremely dangerous. Identification may be doubtful because their colors vary with light, water, soil, and sometimes for reasons concealed by the fungus. Again, mushrooms that are harmless to one person may be poisonous to another, and species that are harmless in one location may be poisonous elsewhere.

Some summers ago, a visiting friend from Indiana picked a half-dozen white mushrooms from a patch outside our log cabin and brought them in to me. When he had gone, I looked closely at his harvest. Five of them were of the species he had intended to pick. They had short, plain stems, and the fluted gills on the underside of their caps were attached to the stalk. The sixth, although also white and grown in the same location, was different. Its very long stem, which was ringed by remnants of a veil, showed traces of a bulb at the bottom, and its gills were detached from the stalk. This was *Amanita verna,* the destroying angel, purveyor of agonizing death. It does not start its symptoms, which resemble those of cholera, until many hours after it is eaten. By this time, the poison has gone into the bloodstream and your physician can only try—and pray.

In the deep shadows, where vegetation rots under the big trees, the Indian pipes grow—gray-white, waxy flowers that look like fungi but are not. Related to the rhododendrons, they lack chlorophyll and, were it not for a special kind of fungus, they could not grow at all. This fungus surrounds the Indian pipe's roots and, taking nourishment from decaying vegetation, feeds the flower.

Such combinations of fungus and root, in which the mycelium of a fungus takes over the functions of absent or feeble root hairs, are called mycorrhizas, which means fungus-roots. They make it possible for evergreens to grow in pure sand, and many forest trees are dependent on them. The mushrooms that are found only at the base of certain trees are the fruiting bodies of their specific mycorrhizal fungi.

Mycorrhizas are responsible for the growth and distribution of many wild orchids. That these flowers grow at all is amazing, because their minute seeds and the microscopic spores of their companion fungi must drift to the same place in a very specialized environment and sprout together.

The calypso orchid sometimes has small, stubby, branched roots below its tuber. Such roots are called coralloid because of their form. Dr. Clyde M. Christensen, of the Department of Plant Pathology, University of Minnesota, writes me that, "of those [orchids] that have coralloid roots, all that have been investigated have a fungus in the interior portion." He thinks it safe to say that the calypso with coralloid roots forms mycorrhizas with some fungus. This would account for the scattering of single calypsos through the thick duff of the woods around the log cabin. It is a unique experience to come on one of these pale-purple miracles of beauty and fragrance, lifting from a heavy brown layer in a place bare of all other low growth.

The now rare pink moccasin flower also forms mycorrhizas. The survivors of this, Minnesota's state flower, left by pickers and transplanters, are under legal protection. It is strange that gardeners, who well know that plants that need sun will fail in shade, expect orchids that are native to duff or sphagnum bogs to flourish in rich, black soil. They may even plant them in dirt that has been treated with fungicide! Our wild orchids are rare. Leave them where they grow.

The business of the fungi is destruction, even when the products of their digestion are used to promote other types of growth. When they attack our food or shoes, or even our feet, in the form of athlete's foot, it may seem that the earth could do nicely without them. But it could not—and remain fertile and populated.

Truly "all flesh is grass," because green plants manufacture not only their own food but also the food that eventually is eaten by all animals. During this process, they absorb carbon dioxide from the air, release oxygen to sustain animal respiration, and use vast quantities of water and minerals from the soil.

In the fall, every fertile part of the earth is covered with a layer of dead vegetable matter, with animal manure and picked carcasses scattered throughout, in which plant food is stored and unavailable. Within and under this layer, seeds wait to sprout and roots to take up nourishment in the spring.

All year the fungi are at work, breaking down the debris and releasing its nutriments for re-use. Without these lowly plants, the earth's resources would have been exhausted long ago and the whole cycle of life ended.

The Flowering Carpet

SOMETIMES, when the winter landscape looks like an etching, with hills and trees limned in needle-sharp black against a skim-milk sky, when the sunlight shimmers white with frost and even my footprints are pooled with ashen shadow, I long for a touch of color.

I find it on a tree where yellow-green moss encircles the trunk with tiny ruffles, or on a stone, blown free of snow and blanketed with emerald feathers. As the snow melts, the mosses brighten into previews of summer green, glittering with gem colors as they catch the sun in the water droplets on their points and branches.

Mosses are very primitive and reproduce by means of two types of plants. A spore generates a branching, green filament body, much like the fungus mycelium. From it spring rootlets and branchlets, with correspondingly small leaflike projections. The supporting filament dies away, leaving the new moss on its own. From miniature organs on this plant come male and female elements that unite, when splashed about by rain, to produce the tiny upright stalk of another type of moss plant. This stalk does not grow on the substratum that supports the first moss form, but digs a foot into the parent plant and takes nourishment from it. At the top of the stalk, which is thin and often bright and stiff like lacquered wire, is a sheathed spore case. The sheath falls away, revealing a capsule with a tight-fitting lid that, when the humidity is right, opens and sheds its spores to resume the cycle.

Mosses grow all around us. An overturned stump is covered with a dark-green mat from which rise half-inch stems, needled like balsam sprouts. Open ground above a subsurface spring, so damp that few plants can survive there, supports a mass of feathery bright green whose fruiting stalks rise like the pile of a thick brown rug. The brook banks are springy with cushions of sphagnum, or peat moss, which was used by the pioneers to chink their cabins and to soften and warm their beds. My Indian friend, Awbutch, when her mother predicted a hard winter some years ago, assured warmth and comfort for her two short-haired dogs by putting peat-moss beds into their houses.

On land belonging to one of our friends is a deep hollow in which lies a small, perfectly oval lake, edged by peat moss that gives under a footstep. This is a quaking bog and, near the water's edge, one must step carefully or break through. In olden times, the bottom of the beautiful hollow was filled by the water of a larger lake, probably with the stony banks so typical here. Peat moss grew thickly around its rim, pushed from the shore outward, and dropped dead vegetable matter which began to fill beneath the floating green. The brown spore cases produced more and more moss, until the dead plants formed the quivering fill that now encircles the water. Eventually the little lake will become a peat swamp and, in some faraway time, will be a fertile meadow, sheltered by the steep walls of the ancient hollow.

Growing near our brook is a prostrate green plant that re-

sembles a lichen but has brown stalks, like the fruiting stalks
of moss, rising from sheaths on its flat leaves. This is a
horned liverwort, a relative of the mosses. Its green body is
made of branching ribbons, much like some forms of sea-
weed. One of these sea ribbons may have washed ashore on
a primordial beach, to send down simple rootlets and become
established as the first true land plant.

Near the log cabin is a granite boulder, dropped by the
receding glaciers. Its several tons are prevented from rolling
toward the house by some cedar trunks. The top of the
boulder, which bears the thinnest layer of soil, is covered
by a dense, gray-green mat. This tough, stiff plant looks like
a moss but does not have the moss's separate spore-bearing
stalks, its fruiting bodies being part of the main plant. This
is the rock spike moss, of another plant family that also com-
prises the ferns, horsetails, club mosses, and quillworts. Our
property has a good representation of all these except the
little-known quillworts, small, clumped plants that look like
chives and grow near, and often submerged in, quiet waters.

Our forest floor shows little green when the snow melts
off, except the land that slopes westward toward the brook.
Here are carpets of evergreen, half-a-foot thick, with some
treelike forms rising from the diminutive groves. These are
club mosses, or lycopodia, often called running or ground
pine and, in some places, princess pine. This last name
creates confusion because the pipsissewa, a flower related to
the pyrolas, is called prince's pine.

The shining club moss grows in patches of bright, enam-
eled green. Its upright single or irregularly branched stems
rise from an underground stem and are covered with pointed
leaves like short needles. Indented rings formed by small
leaves indicate the beginning of each year's growth.

The stiff club moss, that rises from a thinly needled ground stem, is finer and less brilliantly green than the shining club moss, with more regularly disposed branches and smoother ranks of leaves between its annual rings. Many of its upright stems are topped by small, slender cones, which are dull and shattered in spring, fresh and green as new ones rise in summer, and yellow when their spores ripen in the fall.

The tree club moss is as charming as any plant in the forest. Its foot-tall trees branch in sprays or in regular whorls around the main stem. The twig-size branches are thickly furred with acicular green leaves, which are lighter at the branch tips and give a decorated touch. Clusters of oversize, slim cones rise from the "treetops" like Christmas candles. These little "trees" are sometimes sold for holiday decorations.

During much of the summer, ferns and large-leaved aster overshadow the club-moss forest. After the leaves have fallen and crumbled into the duff, I stand ankledeep in the yellow-tipped evergreen, trying to imagine how it might have been to stand, a quarter of a billion years ago, in the shadow of the towering ancestral club mosses that, with tree ferns and giant horsetails, made up the forests that became our beds of coal and jet. Had creatures like me existed then and plodded through the swamps in which the primeval forests thrived, they would have been no larger beside the "trees" than beetles crawling under the club mosses at my feet. And I am just as small against my own forest background.

The horsetails, or equiseta, were among the first plants to establish themselves in the subsoil exposed by bulldozing in the log-cabin clearing a dozen years ago. Their aboveground growth is sparse in dry years, but their perennial underground stems advance through drought, to send up shoots in new places when rainy springs come.

The four-foot, branchless scouring rushes, or rough horsetails, look like small, thin, green bamboo. They were used in early days to scour pans, and it does not come as a surprise that they are harsh to the touch. But it is startling to feel the rough, dry surface of the other horsetails' delicate branches.

The stubby fertile stalk of the field horsetail, topped by a large, brown, cone-shaped, spore-bearing strobilus and banded at each section juncture by a black sheath with sharply toothed upper edge, is very prominent in low, early greenery. It dies after its spores are shed. The twenty-inch sterile stalks are thinner and have narrowed black sheaths, from below which light-green, leafless branches rise in whorls to make a flat-topped form, like a thinly bristled whiskbroom upside down.

Forty-seven stalks of meadow horsetail, like a clump of foot-tall flagpoles, appeared this spring on a piece of ground that is so high and dry that it is almost barren. These stalks branch after shedding their spores, and they and the sterile stalks are ringed with threadlike green branches, spreading horizontally from the stems like the attenuated legs of enormous versions of the harvestmen that I called granddaddy longlegs as a child.

The path leading to the road is edged by wood horsetails, rising two feet tall from the rich, always-moist ditch edge. The upper two-thirds of these enchanting little emerald-green trees are ringed with whorls of dendritic branches, whose growth crosses and recrosses like the threads of needle-made lace, spreading in filigree parasols that droop gracefully at their outer rims. It stuns the imagination to envision them in their ancient form, fifty times as large as they are today.

The ground plants follow the changes from cold to warm to cold again in an irregular parade. The sunny Canadian shore has banks of wild roses, some pink and some white, some single and some double, whose bushes are covered with pale-red hips when our shadowed, pink, single blooms open for their short, perfumed lives. The snow may go off late, so that the plant world burgeons to make up for the shortened growing season. The newly bared ground may be dotted with greenery or frozen out so that the hardiest plants have to sprout anew. The growing season may be cool or warm, wet or dry, and order of appearance, size, even the plants that flourish, vary with these climatic changes. Add to this the wide variety of terrain within our property and generally in this area, and consider the rare plants and the intergrades that show genetic changes at work, and you have a situation that is confusing, full of surprises, and a botanist's paradise. Like the insect world, the plant world by its very size and variety beggars any simple description. But, in 1960, from the time the relatively light snow of the winter was melting, through the rather dry summer that followed, this is what happened.

On the last day of March I was kneeling on a chair, watching the miracle of melting water plink-plonking from the eaves into a puddle, when the sun struck the willow top and I saw the first shining, pearl-gray bud. Around the base of the big pine, where the ground is high and the snow blows thin, were patches of spill-covered earth, the first bare ground of the year.

During the next weeks I explored every day. From under the snow came the gurgle of run-off water and up through the thinning blanket rose the hardiest sprouts. The rhubarb in the garden showed red-and-yellow clumps, and some onions, lost in the shuffle last year, had naturalized and were poking up strong green leaves. There were scattered coarse grasses, and wild strawberry leaves covered the earth west of the log cabin. The ovate leaves of forget-me-nots showed blue-green against the earth and snow. They grow everywhere, all spread from one plant that I set in a high place five years ago. The yellow-green of sweet-william leaves brightened square feet of ground and Johnny-jump-ups had seeded even in the gravel paths, where their little sprouts braced themselves against trickles of icy water. Trailing strands of dark green promised twinflowers, and the scalloped leaves of naked miterwort, bristly as though they needed a shave, were scattered under the evergreens. One by one, widely separated and usually single, the leaves of the calypso, still half-plaited, lay flat against the chilly duff.

The nights were still dropping into the twenties but the thaw swept on during the daylight hours. The buds on the elders grew purple and full, hesitating as late snow fell, proceeding with caution as the days warmed. The birch branches were hazed with buds, and the blue flax sent up its sprouts that look, when they are small, like the first shoots of spruce.

By mid-April the snow was gone but night freezes coated the pools of water with ice, through which a thistle sent up a determined, prickly stalk from the gray-green, cactus-like rosette of its first-year growth. It was wonderful to feel the hardness of the still-frozen mud and the softness of the fallen stems of last year's grasses under my feet. I found a white pine seedling, bent to the ground by the plant debris. Ade cleared a space around the tender little tree and covered it with a wire cage to give it sun-and-air room and to protect it from hungry hares, then went inside to disinfect dozens of little cuts on his fingers. The innocent-appearing grass was sedge, whose identifying three-sided stems should have warned him, and whose sharp-edged leaves, green or dried, cut like stiff paper.

A week later the warm winds of spring soughed around the house, to end the night freezes and start the drying of the sodden earth. The fifty-four-degree air seemed like summer and thunder rumbled tentatively in the distance. The chives shot up and the leaves of late-blooming asters lay like dark-green hearts around their black stems. I filled vases with the pointed-leaved brown sprouts of wild sarsaparilla, which spreads so rapidly that a little discouragement is in order. And now I could see the leaves of violets, yellow-green and close to the earth, around the brown fertile stalks of the equiseta.

May Day passed, not with its traditional flowers, but with young nuthatches and gray jays at the door and the earth flooded with green. Columbines and bluebells, raspberries and wild roses, the multitudinous common plants of the earth—dandelions and plantains and grasses and daisies—grew so rapidly that I could see changes within a few hours, and the green of the equiseta spread mistily.

Two weeks later the giant trilliums were half-a-foot tall. The blue-bead lilies were uncurling brown cylinders and turning them into shiny green leaves. The twisted-stalks and false lily of the valley were well up in the shadows. The bronze leftover bunchberry leaves were fading into the earth as the new green shoots lifted. The climbing false buckwheat was rising beside the cabin wall.

This buckwheat has red stems and dark-green arrow-shaped leaves that, when new, are brushed with red. Three years before, I transplanted several clumps and trained them on wires to shade the window. Now the vines climb vigorously to the roof edge and drape gracefully down, providing not only shade but masses of tiny white flowers and pyramidal green seed pods. Our buckwheat is *not* a weed—not any more.

In sunny places, the silver-fuzzed fiddleheads of the interrupted ferns were already ten inches tall, lovely as any flower before they had opened a frond. And on the day when the maples and aspens leafed and the pin cherries and birches opened their buds, when thousands of jewelweed seeds were putting up pale sprouts and flowers were starting on the elders, the violets bloomed, covering the low-piled carpet with short-stemmed flowers, white and blue and lavender and deep purple.

Now flower stalks rose everywhere and, as though on signal, the flies and mosquitoes buzzed in, ready for the task of fertilization that lay just ahead. On the twenty-fourth of May the buds on the calypso's short stalks opened, each above its flat leaf. These orchids look frail but last a long time. They turn from white, touched with pale lavender and yellow, to deep lavender with a violet crown, orange accents, and purple-brown inner markings that show through the translucent cup. As their perfume fades, so does their color, until they turn white and wither away. Then the almost-round leaf dies, too, and one of the rarest flowers of the forest rests until its little tuber wakes to life under the next winter's snow.

Then there were warm, languid days of rain and sun. While flower stalks readied their buds, ferns grew lush. In sunny places the coarse leaves of the interrupted ferns showed the brown, shrivelled, spore-bearing, central leaflets that give them their name. The graceful plumes of the ostrich ferns circled the stiff, brown, fertile stalks of last season. The six-inch oak ferns lifted their thin, black stems and uncurled fragile triplets of triangular leaves, like yellow-green miniatures of the waist-high bracken. The lacy wood ferns, long-lasting when picked and favorites with florists, sprang up in patches, and the long beech ferns lifted slim, pointed leaflets next to the uncurling tall fronds of the cinnamon ferns. (Both the ostrich and cinnamon fern fiddle-heads are delicious when steamed.)

As I was watering some seeds in a flat, I saw a tiny, prostrate, heart-shaped leaf, sprung from something in the soil itself. Some days later, a thread of green curled up from the notched end of the little leaf and opened into a tiny frond, while the original heart-shaped leaf died away. I had seen a fern begin its life.

The fern, like the moss, is primitive and reproduces by means of two plants. The very small leaf that I had first seen corresponds to the wide-spreading moss plant, and the second leaf was the start of one of the beautiful plants we know. They produce spores, as does the capsulated stalk of the moss, but are not dependent, as in the case of mosses, on the parent plant. The fern's spore-bearing plant in past ages developed the vascular tissue that we call wood, whose tubes convey water and minerals from the soil throughout the plant body.

June arrived with a proper accompaniment of singing birds, sunny skies, and bursting buds. The ground was spattered with the golden-hearted white blooms of the wild strawberries. The forget-me-nots' pink cups opened into blue spangles in the grass. Greenish-yellow blue-bead lilies swayed in the breeze and the purple-pink bells of the twisted-stalk hid modestly under their leaves. The western bluebells, or tall lungwort, dropped waxen sprays, pink to violet to blue as they developed. The globeflowers hung white balls that, when the sun touched them, lifted their heads and opened their petals to the sky and to the searching white-admiral butterflies. These butterflies, not white, but with black wings vertically crossed by wide white bands and edged with rows of mosaic red and blue, hovered over the flowers as though enchanted until the shadows of the big trees touched the petals and turned them into pearly balls again.

Then the log cabin was surrounded by dandelions, my private symbol of the transformation to summer. If sunshine can be caught and distilled and filtered and made safe for mortals to see, the dandelions do it—and there is magic in the energy the human animal gets from their tangy, bitter greens.

The blue-bead lilies did not have long to bloom near the cabin, for our tame doe ate most of them in their first few days. They will come up again from their nipped-off bases and their underground stems, and Mama looked very pretty indeed, standing knee-deep in ferns, her red summer coat bright against a background of maple green, the sun turning her translucent ears a rosy color, as she chewed a bouquet of lilies in a well-bred manner.

As the globeflowers faded, the bunchberries white-starred the duff in cleared places and one-flowered wintergreen and side-bells pyrola dotted the forest floor. The naked miterwort sent up four-inch stalks with tiny greenish flowers and, between each of the five petals, a growth that looked like a patriarchal cross made of stiff spiderweb. Paired twinflower bells hung pink above their creeping stems and white star-flowers looked up from their pointed green collars. Columbine was blooming and, on the edge of the brook, the marsh marigolds opened their golden cups.

On the wet lower bank grew tall bulrushes, and wild blue flags, their equitant leaves bestriding their stalks, lifted a purple bud and an opened fleur-de-lis. Deep in the shade on the higher bank, Labrador tea showed clusters of modest whitish flowers and its leathery leaves curled back as though to protect the white wool on their undersides. Ground juniper, aptly called "shintangle" for the way it tangles one's shins, spread interwoven low branches, which have a changeable look when the wind tosses them because their dark needles are almost white on the underside.

In July, little white flowers fell from the dewberries and raspberries and the big "white-rose" blooms were gone from the thimbleberries. Honeysuckle showed pink and white and orange and yellow on their bushes. Pearly everlasting produced its stiff white balls of petals, and meadow peas and vetchling added touches of cream and purple. Daisies swayed in the grass, which was covered with seed heads—brown and purple, amber and jade. Above their fine-toothed dark-green leaves, bouquets of white yarrow bloomed.

Then the sweet williams, whose variety of color and pattern is endless, came into their own. From a rare, pure-white bloom to a four-inch flower head of the darkest scarlet, from flat single flowers to those so double they look like roses, from cryptograms in red against their white, to white embroidered edges on their red, from deep purple to fluorescent tangerine, the variety grows with the bees' interchanging of their pollen. It does not matter that their blooming points gently toward the last weeks of summer.

August is berry time but, of all the delicious morsels we saw forming on the bushes and hiding under the leaves, we gathered very few. The chipmunks and birds come early for their fruit. I can buy a jar of jam, but there is no price for a *ah!* thrush singing by my window.

Then the goldenrod lifted its torches of yellow and the heart-leaved asters opened their blue-flowered plumes. The burry seeds of the tall buttercup clung to my clothes and the scent of crushed mint followed my footsteps. The sparsely flowered large-leaved aster had spread so widely that I would have a weeding job next spring. And the brown-and-orange jewelweed flowers turned into touch-me-not pods.

I was standing in the moonlight one night when I saw what looked like the specter of a plant. It was a jewelweed, grown to rosebush size, with the pointed ovals of each leaf faintly luminous. These leaves have widely spaced edge serrations, and on each little point a droplet of water glittered in the beam of my flashlight like a diamond. (This plant process of exuding moisture through an uncut surface, which produces such beautiful results, is called, not very beautifully, guttation.)

Indian pipes rose like phantoms through a mat of twigs under a pine. The baneberries outdid themselves—some were white, some red, and one plant produced white berries with scarlet mottlings. A few late daisies bloomed with the black-eyed susans. The pond lilies and spatterdock ceased to float their white and yellow chalices on the still water and the keys whirled down from the maples.

It was September. Gradually the waist-high weeds and grasses went down. Underneath were the plants that had been prominent in spring, but now the strawberry leaves were maroon and the violet leaves were cream-colored. Still upright were the dark-brown fern fronds and the white flowers of the everlasting on stalks draped with blackened leaves. A yarrow emerged into the light and put forth unusual pink blooms.

As the cones seeded on the evergreens and the mushrooms popped up, the hills blazed and the flat blue mirror of the lake reproduced every leaf and branch. Only the blue blooms of the flax were still greeting me in the mornings.

The summer had gone where all summers go, and would not return in the same dress again.

Creatures More or Less
of the Water

AMONG THE DEBRIS that Ade and I cleared away when we moved here was a pile of jack-pine building logs that should have been peeled so that they might dry. They had not been, and wood-boring insects had reduced their damp interiors to the status of firewood. Ade sawed them into stove-length chunks and I took the wheelbarrow to haul them near the cabin, where he could conveniently split them.

As I lifted a chunk, a section of its soggy bark fell away and my fingertips slid into a slippery, gelatinous mass. I dropped the wood, wiped my hands, and gingerly picked up the chunk by the ends to investigate the slime.

It looked like a patch of dark, thickened water, which began to spread by oozing forth rounded projections. Gradually the mass flowed into them and more projections "reached" out, until the slime could be seen to move toward the sheltering damp of the remaining loosely attached bark. This was the animal form of a slime mold, one of the living links between the plant and animal kingdoms.

I carried the piece of wood to a dank and shadowed ditch where the plant-animal would not perish of dehydration. Two days later it gathered itself into a blob on the bare surface of the wood and asserted its vegetable nature by growing a short "grass" of slender, transparent, brown hairs. The next day, these burst and sent up a cloud of spores, and the slime mold's work was done. These spores would release tiny swimming animals that, amoeba-like, would clump together to form another slime mold animal.

I thought of the blurred boundaries between the great divisions of material entities. I recalled electron micrographs that showed a virus enlarging and dividing in the manner of a one-celled animal—but the virus was protein that had been recrystallized several times in the laboratory like an inert chemical compound. No ordinary living thing could survive such repeated solutions and evaporations, yet, in the presence of living tissue, the virus crystals reproduced their kind. And, in the waters of the earth, there are microscopic flagellates, propelling themselves about and reacting to stimuli as animals do, but possessing chlorophyll and making their food in the manner of green plants. Not only is animal life linked to the plant world through creatures like the slime mold and the flagellate, but it is linked to the mineral world through things like the virus.

there are NO blurs!

The natural world offers many wonders, but nothing is more overwhelming than its demonstrations of the unity of all things—animal, vegetable, and mineral.

One would expect a pail of water, dipped from a rock-bottomed lake through a hole in thick ice on a below-zero day, to contain water and little else. I learned better when I looked into my dipper and saw, silhouetted against the white enamel, a cloud of minute animals, swimming in haphazard jerks. Some were red aquatic versions of the spider mites that nibble rose stems. Others, colorless and with propelling appendages almost too frail to be seen by the naked eye, were various miniature crustaceans loosely called water fleas. Busy dots, more like bubbles than living things, may have been the free-swimming larvae of slightly larger crustaceans.

A new water hole, farther out on the ice, gave us less visibly populated drinking water, but water dipped from the old hole near shore showed slender fairylike Cyclops, something like tenth-inch-long shrimps without legs. In a good light it is possible to see the centered eye that gives them their name. There were Daphnia, too, jerking their rounded bodies through the water by means of a pair of feathery antennae. These transparent creatures, sometimes showing faint rainbow colors, are important food for larger water creatures.

After the ice was gone, Ade dipped out a young crayfish, a small replica of the adult, with a white outside skeleton touched with shocking pink. These pretty creatures hatch from eggs attached to the underside of the mother's body and cling there for some time before they swim free in their vast, watery world. We returned the crayfish to the lake and watched it settle on a submerged rock and creep slightly forward. Then, as though responding to a delayed-action alarm, it shot away backwards, as its larger marine cousin, the lobster, does.

All these crustaceans are related to the crabs and to the pill bugs, those oval, shadow-colored wood lice that live under stones and logs. When these land members of a water-dwelling class are exposed to open air, they curl up in a ball to keep their gill-like breathing apparatus moist and operative.

Inland waters do not teem with life as the oceans do, but there is plenty to see along our shore on any calm day. The rocks, coated with a brown fur of mud and plant debris, catch enlarged shadows of the dimples made by the water striders' feet. As these bugs (and, unlike many so-called "bugs," they are correctly named) slide over the surface, sometimes making a little jump toward a drowned fly or

other morsel, the shadows spread and shrink on the rounded
rocks, disappear in a crack and reappear on the upper surface
of a sharp-edged slab. Moving even more gracefully than the
shadows are giant leeches, their ribbon-like black bodies
elongating and narrowing, shortening and widening, flowing
over rocks and twisting into spirals, settling at last in a
hollow where thin sunlight touches a bit of sand. A scattering of heavy-bellied tadpoles drifts above the leech, their
flat tails wriggling. From a crevice, a giant water bug
crawls out and propels its brown body after the tadpoles by
sure movements of its flat hind legs. Its beak snaps up a
tadpole, and the little group of future frogs hurries on, with
furious tail-lashings. A whirligig beetle drops onto the water
and begins its dizzy spinning, the lower half of its divided
eyes looking into the water, the upper half scanning the
air. The water bug turns and swims toward the beetle. It is
almost at the surface when a gull swoops out of nowhere and
the water bug is gone. Through the ripples left by the gull's
splash, a school of little fish, perhaps minnows, perhaps
game-fish fry, moves along with graceful sway of bodies and
fins. Before the water quiets, they have swum out of sight
behind a block of granite. Unless you are a skin-diver, you
get few chances to see fish alive and going about their natural business.

It will surprise some of my Chicago friends to learn that
neither Ade nor I fish, because they assumed that we bought
a home on a lake for that reason. When a visitor asks for
piscatorial advice, we can only say that if one knows the fish
he wants to catch, what it eats, where it may be in location
and depth according to the season, temperature, and rainfall,
he may, if he can handle his gear and has a run of luck,
catch the big one that did not get away.

The mid-May opening of the walleye season brings a rush
of spring fishermen that begins around midnight with a
parade of cars humming up our side road. At about the same

time outboards start up the lake from the public landing at the west end. The unaccustomed feeling of being surrounded by people brought me outdoors at 2:00 A.M. last opening day. I stood in the pitch blackness of a foggy night, breathing in the scents of the forest—and listening to a blend of engine sounds and whining tires, accented by occasional shouts and unintelligible words. If it had not been for the freshness of the air, I might have been standing in a residential district in Chicago during a power failure, listening to traffic on distant boulevards.

When dawn came, I sat on the shore and watched the light sweep the mother-of-pearl mists from a clear turquoise sky. Two boats with almost silent motors were moving east near the center of the lake. "QUIET, ISN'T IT?" rang out from the farther boat. "YEAH. AWFUL QUIET," bellowed the steersman of the second. As the echoes died away, a forty-foot cabin job fit for coastal cruising approached, like a Bentley limousine pressed into service to get a loaf of bread from the corner store. Bouncing on its wake came a fourteen-foot skiff, sunk to the gunwales with the weight of gear and four men, three of whom stopped their tin-can bailing long enough to wave at me with the optimism that is the fisherman's trademark at the day's beginning.

In past years, the opening of the walleye season here has been marred by crowding at a spawning bed in the mouth of a small river that drains into this lake. Boats jammed the water surface and the almost helpless fish were pulled out while they still streamed roe and milt. When the first week end was over, the river mouth was cluttered with a stinking litter of dead and injured fish, beer cans, sandwich wrappers, and fish guts. In the spring of 1962 the wardens closed this spawning bed. If this excellent practice is continued, there should be no need for future stockings of this game fish, for the fish will restock the lake in the old-fashioned way.

It is common gossip that suckers are strong in flavor, bony, and too soft to eat. Actually they are sweet, and the Y-bones are not numerous enough to be bothersome. They are softer than some game fishes, especially during warm weather, but it is noteworthy that many of those who cannot eat soft northern fish sing the praises of the soft-fleshed southern pompano. Suckers are also condemned because trout feed on sucker fry and are said not to be hungry enough to take the fishermen's bait. This does not take into consideration the fact that hungry trout will readily feed on young trout. Suckers make good fish cakes and chowder, can be filleted and fried, or may be canned. They could be an important food fish instead of a headache for the conservation department.

An odd-appearing fresh-water cod, valuable both for its flesh and its liver oil, is the burbot, or eel-pout, that looks half-eel and half-fish. It is also called, for some unfathomable reason, the "lawyer." The catcher of one of these shudders at its snakelike form or snickers sheepishly, as at some discovered indiscretion. Sometimes I think a fish is eaten, not because it is flavorful or nutritious, but because it presents a conventional appearance and has the "experts'" approval.

When hot weather comes, the trout seek the cold water of the depths and can be caught only by deep trolling. This offers considerable frustration, judging from the tangles of broken line that wash up on our shore. (Short lengths make handy clotheslines.) Consequently, many of the summer fishermen are after walleyes, and others go to a nearby lake to catch the savage-toothed, slender northern pike. It seems an unnecessary complication to the art of fishing that the walleyed pike is really a perch, and the northern pike, a muskellunge!

In spring, when the surface water is below fifty degrees, the trout come up to feed on May fly nymphs and other

larvae. One afternoon a rumor started that someone had caught a huge trout near the mouth of our brook, where it had been gobbling minnows from the stream. Although it is well known that our brook is too small and uncertain of flow to produce any small fish, an hour later ten boats had gathered at its mouth. Within an area some fifty by a hundred feet, their occupants wore out the hours from midafternoon to sunset in the intense concentration of meditating prophets.

After they had gone, I sat on a cedar root that overhangs the brook mouth. The sun was setting behind a cloud bank that portended a thunderstorm. As I watched the cloud's edges turn from white to brass to red, a gap opened in the dark mass and a fiery beam penetrated the water below me like a searchlight. A shadow was drifting past, a three-foot, wily trout, moving toward a dragonfly that lay struggling on the gently heaving water. With a flash of silver jaws, the fish disposed of the insect, then hung motionless just below the surface, a prize in the inland-lakes competition that did not intend to be caught. As the sound of the last disappointed fisherman's motor died away, the great trout leaped. Its gray-spotted silver body, as thick as a man's thigh, curved and disappeared in the direction of deep water.

I lingered, watching mosquitoes rising from the water, until a terrified squealing broke out behind me. A garter snake had caught a young toad, which was struggling frantically to pull a hind leg from the snake's mouth.

I seldom interfere in the affairs of wild things but, struck by the pitiful cries, I hurried to the toad's rescue. The snake released its dinner and slipped under some ferns, whence it peered out, hissing and weaving its head from side to side. I picked up the trembling toad, a soft, cool, brown fellow about two inches long, and held it until it quieted and began

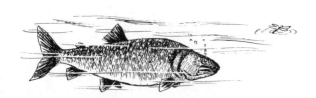

to struggle to escape from me. I left it in some tall grass.

Conflicts between amphibians—represented in the north-eastern tip of Minnesota by toads, frogs, and salamanders—and reptiles—represented by snakes and turtles—have been going on a long time. The amphibians were the first verte-brate animals to crawl out of the sea and maintain a land existence, and the modern descendants of these early ad-venturers, with a few exceptions, still must lay their eggs in water and spend part of their lives there. The reptiles developed the land egg, and from these forward-looking creatures rose the present world of birds and mammals. To-day's reptiles are clothed in scales or plates and the feet, when present, are clawed; the amphibians have smooth, moist skin and no true claws. All of them are cold-blooded, which means that their temperature adjusts to that of their surroundings, and they hibernate during the cold months. The bitter climate of this border country is too rugged for many of them and its deep, rocky waters and coniferous forest supply suitable habitat for only a few species. Unlike birds and mammals, which move when their habitat is de-stroyed, the reptiles and amphibians are not very mobile and suffer great losses when land is cleared or burned. Auto-mobile traffic takes a serious toll of individuals crossing high-ways to reach breeding grounds. It would be regrettable if these ancient creatures should fall victim to expanding civilization.

The northeastern part of Minnesota has only two turtles, the snapping and the painted, both of which prefer sluggish waters with plentiful vegetation. I have not seen any near our cabins and I am sorry. Neighbors from a line that is little changed since the rise of the dinosaurs would be reassuring to a Johnny-come-lately like me.

The snapping turtle is large and, since turtles have no maximum size, keeps on growing through its years. Four

feet long and fifty pounds heavy was not unusual among those that lay on the logs of a shallow Ohio creek near my childhood home. The snapper's shell is very heavy and is roughly knobbed in youth, but may become so thickly coated with algae that it looks like a rock. Its zigzag-scaled tail makes up almost half its length.

On the other hand, the painted turtle is a pretty little thing, dark brown to green, with red and yellow markings on its underside and sometimes along the edges of its upper shell. Its growth is as slow as its spraddle-legged plodding, for it takes almost twenty-five years for a painted turtle to approach a six-inch length.

The snapping turtle's slant-eyed face is equipped with a formidable beak. This beak is not bluff and, unless one knows exactly how to handle snappers, it is an excellent idea to leave them strictly alone, which is all they want anyway. This is particularly true of females that are on land to lay their eggs.

A mother snapping turtle was seen taking care of her family responsibilities on the terrace of a lodge some miles from here. She made several trial diggings with her clawed hind feet before she completed a satisfactory nest cavity. After she had laid her eggs, she was frightened away. Guests at the lodge covered the eggs. When she returned she was killed for her meat. Fortunately young turtles do not need maternal care. If the eggs hatch, the young will be miniatures of the adult snapper, and will dig out in about ninety days. If not interfered with, they will find their way instinctively to the water, in which they will spend the greatest part of their lives.

Appearance is not always a good indication of the class to which an animal belongs, for a turtle and a snake hide their reptilian relationship very adequately within unlike exteriors.

I often see a common garter snake, slender and about three feet long, sunning on the cabin doorstep. It is dark olive-green, or perhaps brown, with three decorative yellow stripes, paralleling the center back and sides. If I stand in the doorway, the snake's head lifts and its forked tongue comes out to sample the air. If I move as though to step outside, it slithers away, its stripes moving with its ripples like gold chains, the sibilance of its movement fading out under the grass.

A snake might be said to walk on its scales. Their unattached rear edges gain purchase against the rough surfaces of stones, earth, and leaves as the snake's limbless body is impelled forward by its intricate musculature. It cannot move backward, and is helpless on a smooth surface like glass. I once found a small garter snake trapped in the open on a skim of new ice that had covered a puddle during a late spring cold snap. Its attempts to move off the ice, onto which it had fallen from an overhanging stump, were steady and persistent until I approached. Then it writhed and twisted, hissing and snapping when I picked it up. It was smooth and dry and cold as the air, and its struggling body left in my hands the feeling of its lean, hard strength. It tried to bite, but it was too small and weak. The strongest garter snake's bite produces only harmless pricks.

In the sandy fill under the flagstones of the summer-house steps is a garter snake's retreat. I found a tangle of young ones nearby in a shady corner last summer. They were from five to six inches long and I had counted forty-one when mother slid into their midst. I backed up before her swaying head, and watched her round up her brood and coil herself around them, still with her head lifted, ready to die for them if necessary. Instinct like that should be respected. I left mother snake's family in peace.

It is likely that the harmless red-bellied snake is here,

but its brown upper coloring blends perfectly with the dead leaves and grass under which it hunts slugs and earthworms, and it is slender and only a foot long. The white underside of its head and its red belly would make identification easy, if I could only find one. It is sometimes mistaken for the copperhead, due solely to confusion of names and colors. The two snakes are very different-looking, the copperhead being much larger, from two to three feet long, with a relatively heavier body that is banded or patterned on the back and sides in shades of buff, brown, rust, or mahogany, depending on the species. The ranges of these two snakes overlap in the south and east, but a brown-and-red snake found anywhere in Minnesota is not a copperhead; this pit viper does not range so far north as the southern tip of the Great Lakes.

Another shy creature that I hope to find someday is the common newt, one of the amphibious salamanders. It inhabits shallow standing water in ancient forest, where there are thick mats of velvet moss, deep duff, and many fallen logs. Our property has some of the proper kind of forest, but the land slopes and the pools drain away. In the mossy earth by the brook, the fallen logs are much too heavy and deeply imbedded to lift in a search for the land form of the newt.

This four-inch fellow starts life as a half-inch tadpole, much like a mosquito wriggler. By the time it is an inch and a half long it is something like a wide-mouthed frog's tadpole. It then metamorphoses into a long-tailed creature, like a baby lizard with four slim, wide-spreading legs. This form

of the newt is called the eft, and its color ranges from bright red to reddish brown to almost black. It replaces its gills with lungs and moves to the land. For three or four years it remains on land, hibernating in the leaf-mold beneath logs during the winter. At the end of this time, it dons a coat of brownish-green above and yellow beneath, irregularly polka-dotted with black and red. Its tail becomes flattened from side to side and it grows fins above and below. As though homesick for its natal element, it returns as an adult newt to the water, breeds, and spends the rest of its life there, moving, even in the coldest weather, underneath the ice, as though driven to give up hibernation for something new.

The spring peeper, an inch-long, brown tree frog with an X on its back, is as elusive as the newt, although its sweet, high, birdlike song rises from the growth around the clearings on moist spring nights. (These and other tree frogs cling to their perches by adhesive toe disks that nonclimbing frogs lack.) No matter how cautiously I try to creep near enough to see it with a flashlight, I find that I am not equipped with silent Indian feet, and the tiny whistler takes alarm, collapses its throat bubble, and remains silent until I am safely out of the way.

Although the swamp tree frog, only a quarter inch longer than the peeper, has been reported in this county, there is some question of its presence in the extreme northeastern part. In the spring of 1962, I heard its loud, metallic trill coming from the brush pile behind the garden. The frog, which may be yellowish-gray or brown or even bright green,

was deep in the tangle and could not be seen, but the call is common and well known through all the United States except the extreme south and far west.

Commonest of our nonclimbing frogs is the leopard frog, about three to four inches long and a mighty hopper for its size. It is a slim-bodied, neat-looking animal. Most of those we see hopping in the grass are bright green, handsomely decorated with irregular white-rimmed black spots. Some have the spots but a more modest brown-green background. Others, not reported previously from this area, are unspotted (except for dots behind each elbow) and their brown skin gives them the appearance of dull-bronze statuettes as they sit under the rhubarb leaves. And one that I woke from its winter sleep when I was raking leaves from a stone wall was brilliant orange, with dark-brown spots widely rimmed with buff. It half-opened one eye and visibly sank back into sleep. I replaced its blanket. Three days later I saw it hopping down a path. It was extremely conspicuous and, since it disappeared soon after, probably made an elaborate lunch for some predator. If its beautiful skin was a mutation, it definitely was not a beneficial one.

The leopard frog's call is a rasping creak, which might be approximated by a novice attempting a tremolo on a bass viol with a poorly rosined bow. Only a lady frog could find it romantic, but to Ade and me it is as pleasant a part of the spring night as the lyrical fluting of the peepers, and as helpful to identification as the calls of the green and mink frogs.

Usually there is a frog sitting half-submerged in a pool by the bridge that crosses the brook. Its wet back is black-green, almost invisible against the dark rocks, and its pale underside offers equally good camouflage from below. It has bright-green lips and a suspicious expression. When the lake level is low enough to partly uncover the rocks along the shore, other frogs sit there, half-submerged and looking very like the one in the brook pool. But the frog in the brook calls for a mate with a repeated *plonk-plonk-plonk,* like the sound of a hammer striking hard wood; it is a mink frog. The frogs at the lake edge grunt in a harsh, exasperated manner; they are green frogs. The three-inch mink frog is only three-quarters as long as the green, and is darker and more spotted, but the size ranges overlap and both species spend most of their time in the water, which hides details. Perhaps there are mink frogs along the shore, but I cannot be sure unless they speak up.

Our finest singer, aside from certain birds, is an American toad whose affinity for insects earned him the name of Flypaper. This species' usual song is a clear, steady trill, sustained for some twenty seconds, but Flypaper is a virtuoso. He sits in the shallow water of our ferny ditch, his expanded throat swelling and shrinking as his trill rises and falls in pitch and volume, until it fades away on a soft, wistful note.

He is exceptionally large, exceeding the usually given four-inch maximum by more than half an inch. He was this size when, on our 1949 vacation, we accidentally dug him out of a burrow. Toads reach full size in two or three years and Flypaper must now be not less than sixteen years old. He may have twenty or more springs left in which to sing sweetly in the ditch. When he came out of his shock at being captured he seemed reconciled to us and, even though we saw him only for brief intervals until we moved here, he became more or less of a pet. At least he does not make frantic efforts to escape if I pick him up. I tried to weigh him, but the kitchen scale seemed a place on which to practice hopping and the attempt was not a success.

There are many toads here, all of the American species. The high grass and thick growth of sweet williams provide shade on hot days, and a subsurface spring makes burrowing for protection easy in time of prolonged drought. The dampness and thick vegetation also provide ample insects for food.

These toads are brown to buff, with thick bodies and warty skin, trimmed by a light stripe down the middle of the back and large, irregular dark spots that often coincide with prominent "warts." Flypaper has greenish spots so arranged that the combination of changing shadows and spots produces a paisley effect as he makes his slow way past the lettuce from which he removes slugs with great efficiency.

A toad's hop is not like a frog's leap, but is a short, laborious heave which ends in a plop, as the soft body spreads comfortably on the ground. The toad's only protection is a disagreeable milky secretion from the skin that makes it distasteful to prospective wild diners. Newly metamorphosed toads are particularly helpless. Once Ade brought one into the kitchen, scooped up on a skin of birchbark because it was much too frail to be handled without injury. It was no larger than my thumbnail, with elfin limbs and a gnome's face, and its very best hopping took it only an inch forward in five attempts!

We deplete many flashlight batteries watching the paths on summer nights that we may avoid crushing the toads, not only the helpless little ones but the large ones that have learned to ignore our passing. With a light, they cannot be missed because their eyes shine like topazes. A toad's eyes are very beautiful, with flecks like gold crystals in their liquid black.

Flypaper is the only one of the path feeders that considers us worth watching. If I stand still, he turns in my direction and looks up. He likes cutworms, which I gather

for him. I drop a worm. His head darts forward, his long tongue flips out from the front edge of his lower lip, the worm is snapped out of the air, and Flypaper's eyes close ecstatically as he ingests his tidbit.

Insects and Other Small Land Things

ONE OF THE EASIEST THINGS TO FORGET when you live in a remote place is that you may not be alone. Had I kept this in mind, I should not have yelled, "Close the door! Do you want this place buggier than it is?" just as Ade was welcoming a lady making her first visit. Her politic murmur about mosquitoes would have saved the situation had not a big olive-green beetle crawled onto the back of her chair and stared at her neck with unfriendly pop-eyes. She hastily departed to tend "something in the oven."

The intruder had bright-green markings on the underside of its thorax and was three inches long, three-quarters of an inch longer than the giant water bugs and an inch longer than a large beetle, whose larvae live under the bark of decaying trees. Our creature was a land form, but its color was far removed from the shiny brown-black of the beetle whose life is linked to wood. We have never seen another and, as our specimen escaped from confinement during the night, it was not identified. Sometimes I think it was created for the chastening of my spirit.

This wild area probably has unrecorded species and subspecies among the multiplicity of insects that creep, crawl, hop, fly, and otherwise disport themselves in the air, on the

ground, through the vegetation, and inside and outside the bodies of other living things. Their numbers and variety are enormous and no season lacks insect activity.

Snow fleas appear on drifts as dark dots, bouncing from one place to another. They accomplish their effortless acrobatics by bending their tails forward and sitting on them, then releasing the tails with a snap, so that they are propelled into the air as from a trampoline. When the sun is warm, a granddaddy longlegs may crawl half-heartedly over the snow or an orange-and-brown tortoise-shell butterfly wake too soon, to flutter like a leftover autumn leaf on sunlit wings, until darkness and the night's deep cold end its futile search for food and a mate.

Our only permanent house guest is a small, dispirited-looking, dusty-brown beetle, with a dingy yellowish band across the forepart of its back. This is the larder beetle, whose plump, hairy, half-inch larva eats any kind of dried animal matter from cheese to hair. How these creatures survive is a mystery, as our food is protected and we see no signs of nibbled leather or felt. Occasionally I find a short tunnel bored into the time-softened paper of my shelved copies of *Ellery Queen's Mystery Magazine*. Inside is a cream-white pupa that has plenty of spirit for a supposedly inactive creature. When its retirement chamber is exposed, it huddles close to the inner end. If touched, it flips its rear so vigorously that it jumps off the page and falls to the floor, where its helpless body bounces about—flip-flip-flip—as though exasperated. I replace the frisky creatures in their tunnels and keep watch, hoping to see the metamorphosis completed, but so far they have evaded me as though by intent.

In early spring, some winged thing drifts across the living room, to crawl sluggishly on a windowpane—a house fly, a mosquito, a "sour gnat." The last is the fruit fly used widely in studies of heredity and mutation. It, like all true flies, has only one pair of wings and does not grow in the adult stage. Whatever the early insect visitor, we welcome it as a herald of summer, reminding us that soon, when the snow is gone, we will scoop half-inch black carpenter ants from our floor and toss them outside. Their restless invasion of the cabin sets us watching for their marriage flight.

The newly matured males and females rise from the nests in logs and stumps by hundreds, mica wings and jet bodies stirring the air into glistening life. For hours they circle and mate in the air. Then, their instinctive task accomplished, they settle, the males to die soon after. Discarded wings turn end over end as they drop from the eaves—frail, veined transparencies developed through the ages for this one flight. Forgetful of their moment in the sun, the females scurry around, examining rotten logs and insect-damaged tree roots, where each can seclude herself to lay her eggs and try to start a new ant colony. Not many survive. Chipmunks feast on them, and gray and blue jays swoop down with the regularity of pendulums. Many find no suitable nesting cell and perish. Sometimes a whole flight is lost, as was one that came from a nest on the lake shore, flew over the water and fell into it, one by one. The strong, the lucky, are few, but from them come the generations of the future.

These ants enter trees through scars and nest in soft, moist wood, such as fungus-weakened pine heartwood or the lower heartwood of white cedars growing in swamps. Because they do not attack wood that is not softened, and tunnel only enough to accommodate the colony and the organic matter that they bring as food from the outside, they do not cause great damage to timber trees. Ants that nest in the earth are usually beneficial.

On the south side of the log cabin, the wide eaves protect the earth from direct rain and the ground slopes so that such spatterings as reach the soil drain away. When we

moved in, we had a sunbaked bit of dusty desert. The
toughest wild plants could not survive there and Ade covered the eyesore with close-spaced flat stones.

Straightaway a colony of small brown ants moved in,
making a main entrance between two stones thicker than
the others and opening doorways in other places as they
tunneled. Windblown rain entered their burrows in bad
storms and the earth began to hold some of the moisture.
Spindly grass and a dandelion sprouted and grew. The soil
darkened and earthworm castings appeared. Now the
spaces between the stones support a self-seeded bed of sweet
williams, far more beautiful than anything I might have
planted.

Once I spilled coconut on the kitchen floor and the little
brown ants formed a queue, coming and going in single file,
carrying every crumb home. Occasionally I deliver a spoon-
ful to their doorway. It is interesting to watch their sys-
tematic removal of this unforestlike food, and small enough
return for the flower bed they helped cultivate.

Earthworms are not related to the crawling larvae of the
insect world, but are segmented worms of the phylum An-
nelida, which means arranged in rings. They are an outstanding example of simple creatures that greatly affect our
lives. They ventilate and cultivate the ground. They fertilize
it by dragging leaves beneath its surface. They swallow soil,
which, after digesting its organic matter, they expel on the
surface as castings, rich with the by-products of their
metabolism. This is fine cover for seeds that might not
germinate otherwise. The earthworm's agricultural importance is so enormous that much of the plant and animal life
on earth might not long survive without them.

The value of the earthworm has been known from the
time of Darwin but, even today, little is understood about
the effect of other lowly plants and animals of the earth on

our food and, in turn, our health and future. For a long time it was thought that soil became "exhausted," in the sense that the supply of required nutriments was used up by repeated cultivation of one crop. It is now known that failures of certain crops in fields so misused are due to specific fungi that parasitize the roots of the plants. With each replanting of the crop, the fungi increase until the accumulation reaches a destructive level. Today's sterilization and chemical poisoning of soil take no account of the known services of small biota, or of the possibility that organisms vital to man but whose values are not yet known may die along with harmful things. Even "harmful" is a doubtful term, because organisms that appear so may be necessary to the balance that keeps soil fertile.

The garden slugs that have such voracious appetites for one's best lettuce are snails without shells. Because of their naked appearance, their liking for moist earth, and their sticky protective coating, they are sometimes thought to be relatives of the earthworm. However, both slugs and snails belong to the phylum Mollusca, meaning soft-bodied, and are related to oysters and squids.

ugh!

At the side of our log cabin's doorstep there is an opening into the earth, above which the gray granite of the foundation is padded by nile-green moss. In damp weather, moisture pearls its minute branchlets and spreads over the stone, bringing glitter to its crystals and touching the red sand in the joining concrete to a muted plum color. Here, under the shelter of a fern's pale fronds, our land snails rasp bits of green alga from the stones in the night. Their flattened, half-inch, coiled shells are marked with brownish-purple that, in torchlight, seems an extension of the colors of the wall to which they cling. The dozen adult snails are the hermaphroditic parents of the eighth-inch miniatures that

wear shells like chips of shadowed pearl and feel their way with sensory antennae like fine black threads. Slowly, slowly, each snail extends its adhesive, flexible foot; slowly, slowly, it moves, leaving a shining trail.

When Flypaper, our toad, hops onto the step, he ignores the snails. Perhaps he has had his fill of slugs in the lettuce; perhaps he does not care for packaged food. He definitely dislikes spiders, ducking his head to the ground when one drops out of the night on a rope of silk. He may have sampled one and had his tongue bitten.

Spiders are not insects, but belong to the class of invertebrates called Arachnida, which means spider, and is represented in this region by many spiders, by a few small mites, and by harvestmen, of which granddaddy longlegs is one. Insects have three pairs of walking legs, often have wings, and have only two compound eyes. Spiders have four pairs of walking legs, never have wings, and usually have eight simple eyes.

Sometimes in early spring, spiders appear in the house from egg sacs tucked into cracks in the logs. They are dark and small-bodied, with a blurred design of dusky red and black, and grow to a maximum legspread of about three-quarters of an inch, molting several times as they become too large for their current outside skeleton. We scoop them up with sheets of paper—not because their bite is dangerous, but because this is the easiest way of getting them off the rugs alive—and toss them onto the step outside, where they are often snatched up instantly by gray jays. Usually one or two find suitable corners and weave flat webs with funnels leading to hidden chambers, where the occupants wait for their prey.

In 1960, one of these webs was in the corner of the kitchen windowsill. It was a dry season with almost no insects. Each day the spider strengthened her snare and waited patiently in the funnel opening for food that did not come. One evening a moth flew against a hot lamp chimney and singed its wings. I dropped it into the web and the spider rushed out to sting it and carry it into her chamber. At the end of a week's feeding, the spider came out when I shadowed the web with my hand and, in another week, she accepted living flies from my fingers.

Dew, or a light drizzle, turns the webs of these funnel weavers and of the sheet-web weavers into jeweled carpets on the grass, and strings the latticed circle of the orb-weaver with crystal balls. The orb webs, with their strongly secured radial supports and evenly spaced spiral snares, are as attractive as their spinners. The large yellow or orange-and-black garden spiders are of this type and one species here is the gray of old bark, with a lustrous, pale, abdominal design that mimics a small lichen and makes the spider almost invisible against the forest background.

One of these camouflaged "engineers" constructed a most unusual web support in an arbor near our summer house. She rode down from a crosspiece on a small spruce cone which was attached to the end of her thread. After strengthening this thread with several thicknesses of silk, she attached the cone to the crossbeam above by two more multiple-stranded threads, slanting outward and upward from the cone to form a triangle, topped by the crossbeam and with the cone at the bottom apex. Inside this framework, the spider spun her food-catching orb web. The whole construction swayed in the wind but offered too little resistance to be damaged by air currents.

The little gray-and-black jumping spiders hop on our forest logs and stones as they do on city windowsills, and white or yellow or pale-green crab spiders with extra-long front legs and a habit of sidling lie in wait for insects on flowers. It is hard to say whether the flower or its matching spider is the more beautiful.

In startling size contrast to these are the huge fisher spiders that catch tiny fish and tadpoles occasionally to enliven their basic diet of large shore and aquatic insects. Ade and I watched one of these as we sat on the summer-house dock. It was light gray, finely patterned with black, and tip-toed across the quiet water like a ballerina, its slight weight upheld by surface tension. Then it dropped flat on the surface, paired its fore and hind legs on both sides, and moved ahead with a breast stroke. It stepped onto the lower portion of the dock where we were able to measure it accurately by the wood grain. Its body was one-and-a-half inches long and its legs spread five inches, which may well be a record for this area, comparable to the size of the female fisher spider of the Okefenokee Swamp and other southern locations. These vie with the wolf spiders for being the United States' largest "true spiders," a term applied to the members of the suborder that includes most of our common spiders. The tarantulas of the southwest are larger, but belong to a different suborder.

I sometimes wonder what would happen to warm-blooded animals if the spiders and forest insects that feed on biting winged things were suddenly to perish. Early-spring mosquito bites are angry-looking and extremely itchy, but the irritation from later bites is less. As our mosquitoes are not malaria transmitters, I have experimented by allowing them to bite me and I believe that their irritant may set up a temporary immunity to its effects. Only female mosquitoes take blood as part of their diet; the males content themselves with plant juices. The little humpbacked black fly injects an anesthetic and anticoagulant, so that one sometimes discovers the bites only by a trickle of blood. A few hours later, these

bites raise welts like bold hives on Ade, and produce inch-
wide purple circles on me. The deer fly's bite is a fiery sting
that draws blood. When one is trying to weed the garden,
these and the black flies rise from the ground in such clouds
that the air seems darkened and every breath tends to draw
them into the nose and mouth. Midges that bite are called
"no-see-ums" because one feels the sting before one sees the
tiny biter, which is only a twenty-fifth of an inch long, with
transparent wings and slim, black body. The bite leaves no
after-effect except a feeling of incredulity that any creature
so delicate could penetrate thick human skin.

None of these bites is dangerous unless it becomes infected, usually from scratching, which only increases the stay
of the irritation. My remedy comes from a homesteader, to
whom I am very grateful. "Soak the itches in hot water—as
hot as you can stand without burning yourself—until the
sting goes out." This really works and is a blessing for bites
on the hands and feet, where use of the fingers or friction
from shoes can produce dreadful itching.

hot water takes sting out mosquito bites

All of these pestiferous creatures are more numerous in
wet years, because they pupate in water or moist ground. Yet
wet years are the years of lush growth, of renewed strength
for trees, of deep filling of the lakes, of freedom from the
danger of fire. And in such years the need for these insects
as pollinators is greater than in dry years of poor growth.

Many beautiful and harmless creatures depend on them,
or on their larvae and eggs, for much of their food—warblers
and woodpeckers, bats and shrews, fish and dragonflies.

Almost as fast as the mosquitoes rise from the still waters
of evening, the dragonflies take them, now hovering in one
spot, now rising abruptly, now swooping almost to touch the
lake surface. These jeweled, lacy-winged helicopters of the
natural world are busy throughout the day and I often see
dozens of them overhead, sometimes accompanied by the
metallic-red ruby-spot damselflies, always moving, snatching

the small winged things out of the air. One afternoon Ade and I were besieged by a whining cloud of mosquitoes. A soft roar joined the whine as four dragonflies darted in. Within two minutes, the "dragons" had swallowed most of the mosquitoes and moved to the assistance of our tame doe, standing with head down, nostrils pinched, and eyes partly closed as she switched her tail and flapped her ears and stomped to rid herself of the biters.

Clearing out a substantial portion of our small trees and brush to permit more air movement would eliminate many of our bites—but the trees and brush are our winter windbreak. Mowing low growth that harbors insects would eliminate more of them—but it would also do away with our flowers and ferns, our berries and horsetails. And with the cut-out growth would go not only mosquitoes and flies, but all things that shelter in it. Spraying with herbicide we do not consider, not only because of its potential danger to health, but because it is not selective in its destruction of life.

Ade and I came here as transients, to see the forest, its life, and its progression. We have no desire to play God to a world so complex that in a lifetime no one can more than glimpse its working. The "bug" season is unpleasant for only a few short weeks in early summer and we take it as it comes.

A hot afternoon in the woods is pine- and balsam-scented, cooled by the shade and moisture under the trees, but it is still hot, especially when one has been cooking on a wood range. On one such occasion I tried vainly to soap away the sticky feeling and ended by dousing liberally with cologne. Feeling fresh and smelling elegant, I walked along the path, enjoying the warm breeze.

Suddenly I was besieged by bumblebees and I thought, as I withdrew as calmly as one can under such circumstances, that I had disturbed their nest. But the bees did not seem angry, although they buzzed back and forth and bumped against my face and neck. Then I realized that I was redolent of the scent of pink clover and the bees were trying to find the flowers.

Bumblebee colonies do not winter over in nests as do honeybees, which are not native here, probably because of the long, severe winter. In the fall the young bumblebee queens mate, then seclude themselves in protected retreats, sometimes in the ground. The rest of the colony dies with the fading of warm weather. In the spring, each surviving queen finds a home in a deserted mouse nest or other ready-made hollow, lays her eggs in a cell supplied with honey and pollen, and seals the cell to start a new colony. The bumblebees that emerge early in the warm season are the fuzzy, black-and-yellow female workers that enliven the garden all summer, from the time of the first white-clover blooms. They do not mate, and the males and queens appear later. The workers are so fond of the honey-scented purple thistle blooms that they cling as though drugged while I touch the soft yellow hairs of their jackets.

I stood with a visitor watching one of them dying on its flower, trying weakly to gather more pollen. This bee, slipping into oblivion with its pollen pockets full, had completed its appointed days and was quietly leaving its world. It is regrettable that man's last days cannot be so well spent and simply ended. But my friend thought it very sad that the poor bee could not take its pollen home.

This is the kind of maudlin sentimentality that has made "sentimental" a derogatory term in connection with wildlife. Although "sentimental" may imply emotionalism so excessive or affected that it is silly, any thought influenced by feeling is sentimental in some degree. Every time I remember the trust of some departed wild friend, I feel its loss and hope that its end was quick and not too filled with terror and pain. There is nothing maudlin about this; I am simply not calloused to the suffering of other living things. Nor do I want to be.

It takes self-control not to swat at large buzzing creatures that hang around one's head, but it is the only safe course. The "buzzer" might be a harmless hawk moth, but it might be a wasp of one of the types called hornets or yellow jackets, whose stings are formidable. They are large, smooth insects, often black and yellow and, in one case, black with abdominal markings of dull white. One of these hornets got inside our screened porch and Ade, who was unfamiliar with hornets, tried to scoop it up with a cloth. Fortunately the cloth was thick. Even so, the slight prick of the stinger sent excruciating pain from his thumb to his shoulder. Fly swatters and sprays are dangerously uncertain. The foolproof way to get any insect out of the house is to cover it with a tumbler and slide a sheet of thin, stiff cardboard between the rim of the glass and the flat surface you have trapped the invader on. (The tumbler makes a good observation chamber, too.) Set the covered glass upright outside and remove the cardboard, so that the prisoner may depart under its own power. And come straight back inside, if it is a hornet you are ejecting.

Defoliating insects of various kinds are ordinarily present in the forest in unimportant numbers, but, when weather and forest conditions are favorable or some biotic change causes unbalance, they may appear in a devastating horde. Such an upsurge of spruce budworms occurred in the northern coniferous forests in 1912. It is significant that, in 1909 and 1910, late spring freezes in Wisconsin, Iowa, and southern Minnesota had destroyed thousands of migrating birds, thus reducing the number of insect-eaters nesting in the conifers.

A budworm infestation appeared here in June, 1956, following dry, warm weather, optimum to the insects' development. The worms first ate the new growth on the balsam firs, then moved to the spruces, leaving the twigs bare or covered with a brown residue when they pupated. The moths, of a gray, mosaic wing-pattern, emerged soon after to lay their eggs on balsam needles. The eggs hatched; the minute worms wintered in silk hibernacula and began feeding when the new balsam growth opened the following June. After seriously defoliating our balsams and white and black spruces during the springs of 1957 through 1960, the budworm numbers decreased around our lake in 1961. Heavy spring rains in 1962 so far reduced the infestation that very few moths emerged from the scattering of undersized pupa cases to lay eggs for 1963's hatching of larvae. I have to look closely to find traces of the 1962 damage amid the scars left from previous years, and the affected evergreens are lifting new tops and producing cones for the first time since the rise of the infestation. A letter from the U.S. Forest Service tells me that 98.7 per cent of the budworm population must perish before egg-laying if there is to be no increase the following year. If that happened in 1962—and continues in succeeding years—we will avoid DDT spraying which deters but does not eliminate budworms and has harmful side effects.

The rise of infestations and the prominence of the biting and stinging insects too often reduces the North Woods insect situation to "Aren't the bugs awful?" This is a pity, because any warm rainless day offers an interesting sampling of insects and other small creatures that will be unlike that of the days before or after.

Today a dobson fly, with lacy wings more than two inches long, is clinging to the stem of an alder by the shore. Its aquatic larva is the hellgrammite, so much prized by bass and other fish as food and by fishermen as bait. I hear the

shrill "song" of a male cicada, calling for a mate by vibrating
membranes stretched within sound chambers on the underside of his thorax. He is an adult seventeen-year "locust" and,
when he burrowed into the ground to begin his long matura-
tion, Ade was an electrician on a Navy tanker off the coast
of Japan. A big, blue-green carrion fly circles my head, but I
offer no suitable substratum for her eggs. Her larvae will remove the decaying flesh from the bones of some dead creature. I hear a soft, muffled *crunch-crunch* from the trunk of
a dead aspen, where fine chips fall as a borer helps return to
the earth the plant food tied up in the wood. A leafhopper
is reducing a sarsaparilla leaf to a transparent grid of veins.
It is not the ordinary green or brown "hopper," but one with
yellow head and wings marked with red and green. Not far
away, on a wild rose stem, blobs of whipped-up saliva protect
the leafhopper's soft-bodied, sap-sucking infant. Across a
beam of sunlight, strands of silk drift on the breeze, carrying
adventurous spiders to new worlds. The silk gleams with
metallic colors, this strand bronze, this one blue, others red
and gold and green. In a maple leaf is a serpentine mine,
made by a small larva that spends its early days between the
upper and lower "skins" of the leaf.

On a willow twig clings the split and empty pupa case, a
baroque, light-brown thing, from which a mourning-cloak
butterfly has flown away on the most beautiful wings in the
forest. Deep brownish-purple are these wings, sometimes almost black, but always velvety and changeable. The borders
are marked with pale yellow, as though edged with age-
darkened fine lace. Between the lace and the velvet is a band
of jet black, embroidered with sapphires, so iridescent that
they shine like moonlight through stained glass. If this is
mourning dress, it does not commemorate any ordinary loss.

Perhaps the mourning cloak honors the passing of the Great God Pan.

The clouds of spring azures, blue as flax petals or silvery as the dawn sky or brushed with the soft shades of wood violets, come when the snow is almost gone. They turn our compost heap into a flower bed, alive with the lifting and falling of their wings. Later the tiger swallowtails drift from forget-me-not to clover to wild rose, and finally replace the blues on the compost. They cover it with triangles, pale yellow striped with midnight black, set with red and blue jewels in the apex formed by their tailtips.

There are so many butterflies—white, sulphur, gray, brown, black-and-white—that it is hard for anyone but a lepidopterist to identify them. Although moths usually fly at night, some join the butterflies in the daytime and add the difficulty of separating these two suborders. Butterflies always have club-shaped antennae with knobs on the end. Moths, except for a few tropical rarities, have threadlike or feathery antennae, or antennae with knobs that are near, but not at, the ends.

The moths rest with silver-dusted wings against the screen beside me, their eyes reflecting pink and yellow from the tantalizing light of my lamp. They flutter in alarm as a horny June beetle bumps against the screen. Beyond, the summer night hums and sings as with the tuning of a thousand distant violins as the night mosquitoes rise. Fireflies dart through the branches like misplaced meteorites and there is a rustle in the ground leaves, as some small thing passes by. The little moths stir and the furry body of a luna quivers

against the screen. Its wide-spreading wings are like pale-green gauze, drawn down into long and graceful tails. The front edges of its forewings are edged with purple and in each wing is a transparent spot, rimmed with yellow, blue, and black. Its antennae lift like gold feathers. I watch this wonder for a while, then blow out the lamp. The luna is too frail for my coarse human world. As the room darkens, it goes back into the mysterious night.

Canteen for Forest Dwellers

WHEN Ade and I moved here, we expected to see birds and mammals near the cabins because the place had been vacant for six years except for our vacation visits. Instead, the red squirrel's nest in a tree near the log house was hastily vacated and the muddy lick by the brook stayed clear of tracks. In an area where people were not known, curiosity might have brought some of the wild things near, but they had developed a healthy sense of caution in this summer-tourist country.

We moved slowly, spoke quietly, disturbed nothing that we did not have to move. If a hare appeared, we froze and let it hop away. If a blue jay screamed at us from a branch, we walked on as though we did not know it was there. The tracks appeared again in the mud of the lick and we saw furry flashes through the trees and grass.

Having established ourselves as reasonably harmless, it was up to us to make our immediate surroundings attractive in the one way that man can show his good intentions to wild creatures. We must put out food. However, considerations such as getting in our supplies and wood took top priority and an accident that left us without a car changed the whole picture.

The windblown snow was deep when we took stock of our feeding-station supplies, which could be augmented little because all winter-bought goods had to travel three miles from our mailbox in Ade's packsack. We had only suet mailed from town, cracked corn bought for our few laying hens, and graham crackers salvaged after falling into the woodbox. Not too hopefully, we spread our offerings on a bench under a white cedar tree.

Within minutes four whiskey jacks soared in on gray wings to gobble suet and carry away crackers. (Whiskey jack is the local name for the gray or Canada jay, which I have used throughout this chapter to avoid confusion with "blue jay.") Soon a red squirrel scouted from a branch and scurried down the trunk to scatter the birds and sample everything. A pair of blue jays examined the surroundings most carefully from distant trees, and downy and hairy woodpeckers appeared, and peeked around tree trunks at the cabin while waiting their turn at the suet. The next day the blue jays came down to the corn. Chickadees and nut-hatches, feeding on the opposite side of the cabin, located the food a week later. Our wilderness canteen was going to be a riotous success, and the riot was not altogether figurative.

With the whiskey jacks competing as families and the others as individuals, the hubbub resembled that around a bargain table. Suddenly a pileated woodpecker dropped onto the bench, which was instantly vacated by all the smaller guests except one nuthatch that clung to the edge, wings spread, beak open, hissing and swaying like the front half of a small dragon. After the big red-crested visitor had sampled the suet and flown away, the others resumed their squabbling.

The chickadees gave way to the nuthatches that yielded to the whiskey jacks that avoided the blue jays that stood about even with the hairy woodpeckers. All the birds flew from the squirrels except one hairy, that hopped up behind any squirrel not on guard and rapped it forcefully in the rear. We did not intend to restrict our canteen's clientele, so we decided to keep the peace by adjusting feeding methods to separate the guests.

Today the bench still stands under the cedar tree, but the station covers a thirty-by-fifteen-foot area outside our door. We serve corn and crackers on flat wooden trays with low rims that minimize spillage. We place these six or seven feet high on tree boles, within easy reach for filling and cleaning. They are popular with certain squirrels that like to survey their surroundings in a lordly manner while eating, and with the blue jays, who are continually looking for danger. As many as ten chipping sparrows have fed together in one of the trays, the birds arguing about the crowded conditions but eating heartily nevertheless. A lower shelf by the door, protected by the ends of the wall logs and sheltered by the eaves, pleases the flying squirrels. They come nightly for corn, scraps of fat, and graham crackers.

We distribute cracked corn in small piles—near the grass-hidden nests of the meadow mice, at widely spaced intervals beside our paths so that the red squirrels can feed with only admonitory chattering, and in any sheltered place for the timid jumping and deer mice that feed in the absence of weasels.

To prevent the squirrels' carrying all the suet home, we at first nailed it to the bench in wire-mesh cages. When a fisher, that big, dark cousin of the weasel, removed the nails from one of the cages and turned it back neatly to empty it, Ade devised a container that could be taken in at night. The cage, made of half-inch wire mesh, is eight inches long, with mesh bottom and hinged mesh top, fastened by a twist of wire. It is stapled to the lower portion of a four-by-ten-inch

wooden backing, and the whole is hung over a nail on a tree by a hole in the top of the backing. Once a forgotten container was removed during a winter night by a fisher. The next spring Ade found it two hundred feet away under a brush pile, its fastener carefully untwisted and its contents deftly clawed out.

Gradually the birds and mammals learned to continue feeding as we moved about the yard, and it was time to try hand feeding. This requires, at the start at least, relaxation, a complete lack of apprehension, and the ability to sit or stand with the steadiness of a student of yoga. Once your customers have become used to accepting food from your hand, reasonable movements do not frighten them. Ade often comes in with stovewood in his arms and a squirrel riding his shoulder, and the chickadees settle all over his shirt and cap when he is going about some outside chore.

The whiskey jacks were the first to come to our fingers, with the black-capped chickadees a close second. The brown-capped chickadees have been slower and prefer our tossing bits of cracker on the ground for them. The blue jays keep a wary eye on the proceedings and will fly down for food. They even call from the tree by the door in pleasant, conversational twitters, but they have not yet lighted on our fingers. The woodpeckers occasionally look interested in suet held in our hands, but they have not approached from the branches. Perhaps this is best, considering the whacking power of their beaks.

Both red and flying squirrels feed readily from our hands throughout the year, the former by day and the latter by night. They tend to be nervous and it is important that the food—in our case the ever-welcome graham cracker—be held so that it extends beyond the fingers to minimize mistakes in distinguishing fingers from food. The red squirrels especially become very excited when several of them are begging at the same time, because it is their nature to compete as individuals. I am always alarmed when friends try to handfeed them, moving the crackers, making strange sounds, while the agitated squirrels leap frantically at the hands. No one has been bitten but it could happen and, even though the bite of these small fellows is not severe and their food is clean, infection might follow.

It is said that tame squirrels will bite an empty hand because they are angry when no food is offered to them. The fallacy of attributing human methods of reasoning to squirrels is apparent here. Red squirrels do not see very clearly. If I throw a piece of cracker on the ground near a begging squirrel, he (or she) may not be able to find it even after considerable sniffing and searching. If I offer him empty fingers, he will reach out and try to take a finger away with his teeth. When it does not come off, he will either jump back in alarm or dig his teeth in as he tries to pull harder. All he wants is something to eat. Any biting would be a side issue, and my own fault for teasing the squirrel.

Our two night-feeding red squirrels have made a remarkable adaptation. They nest a city-block away near the brook and are repulsed violently during their normal daylight feeding hours by the squirrels that live near the cabin. After it is dark, they slip in quietly, scratch at the screen and wait, one inside a pail hanging in the woodshed and the other in a hole under the step. They come out when I open the door, take their graham crackers gently, and hurry home, sometimes making three or four trips. After storing the crackers, they return to eat corn that I put close to the foundation so that they will have some protection from owls. We are particular about feeding these two visitors because their trip through the night is a long and hazardous one.

Wintertime, when food is scarce, brings the largest group

to our feeders. There is no cautious hesitation then. When we open the door to the dusk of a white morning, the motionless snow world comes to life. Red squirrels scamper to our shoulders and stand tall at our feet, begging for new piles of corn to replace those that are covered with snow, although they will dig up the buried food if necessary. Woodpeckers hop up and down the tree trunks while Ade hangs the suet containers. The ever-hungry whiskey jacks glide from the treetops. Chickadees and nuthatches cling upside-down and sideways on the trees until the woodpeckers leave the suet. One nuthatch, maybe a descendant of the one that dared the pileated, is brave enough to feed with the woodpeckers. And the blue jays add glamor, dropping from snowy balsam boughs to flash their violet-and-turquoise plumage against the yellow of the corn.

Snowshoe hares nibble grain left over from the feeding of the other animals. In summer they find our carrots simply delicious! While Ade was pounding in posts for a fence to keep hares out of the garden, one of them sat casually in the path, eating dandelion leaves and watching him with interest.

Deer feed on grain and cedar branches that Ade cuts with a pruning hook and sets upright in the snow. The first of these was Peter, who, near starvation, walked out of the woods on Christmas afternoon in 1958. He chewed desperately at dried thistle tops and raspberry canes, and even crunched fragments from turkey bones tacked on trees. I fixed him a handsome dinner—potato skins and carrot tops, cracked corn and wheat cereal, a ration of suet, and some egg shells, the whole liberally salted. No sooner had I set this on the snow than a red squirrel settled to enjoy the feast. The buck, big and hungry as he was, finally licked cautiously at the twitching tail of the squirrel, whose claws could scratch a nose painfully and might even destroy an

eye. The squirrel did not budge. Peter stared wistfully in the window, his pink tonguetip out like a beacon of hunger, until I went out and chased the squirrel. In the morning the buck was waiting under the snow-frosted branches outside the door. He stayed on with us, sleeping near the cabin. Next April, when he left for the summer, he was fat and sleek.

Weasels come for handouts when the mouse crop from the previous summer is somewhat depleted. Our standard for them is ground beef. The birds, after some time, learned to continue feeding while I am feeding a weasel, although they never lose sight of the little carnivore on the doorstep. I have even managed to feed a weasel at my feet with one hand and a whiskey jack with the other, held high above my head. The adjustment of these natural enemies seems to depend on me and my food gifts as a buffer. When I go inside, the truce is over and the birds renew their efforts to drive the weasel away.

Hunger brings fishers to us in cold weather, too. They are shy but friendly and behave much like their smaller cousins, the weasels, gliding back and forth on the doorstep to be handfed. One winter a frosty old male that had lost a forepaw in a trap joined our night-feeding club. No amount of coaxing could persuade him that we were harmless, so we left meat scraps for him in a pan under a large carton, which prevented the flying squirrels from stealing his dinner. He lifted the box, cleaned the pan, and thereafter, for reasons known only to himself, always turned it upside-down before he left.

So many of the wild things retire to tend families in early spring that the yard is relatively empty. However, red squirrel mothers-to-be continue to feed with us after their late-March mating time. Though they chatter warnings to each other, they seem to know that there is enough for all and do not fight seriously for the territory that includes the feeding yard. And in April the chipmunks, still fat from the corn we gave them to store the previous fall, patter out over the melting snow.

Soon it is sparrow and junco time again and, almost before we realize it, summer, the time of young things. Whiskey jack and blue jay children flutter in the branches and drop down to feed, imitating their parents' voices and behavior. Baby chickadees and nuthatches accept suet from their mothers' beaks, and a young male downy woodpecker, his scarlet headpatch a bit blurry, his vest a soft gray, was brought by his mother to our suet cages last summer.

In early June, the red squirrel mothers disappear for a day or two. They return for hasty feedings until they introduce their little ones at about two months old. These youngsters peer at us with intense curiosity and come by little hops and jumps to examine us closely. I know of no better cure for the human doldrums than to sit quietly, holding out a piece of graham cracker to a very new, very fresh-looking, very well-groomed baby squirrel, until it at last stretches out to take its first bit of food from your hand.

Crows stalk pompously through the grass looking for scattered corn. Gulls circle down when Ade visits the shore. A raven, attracted by meat in a doubtful condition, spreads consternation among shrews already enjoying this banquet.

Suddenly the aspen leaves are gold. The red squirrels are quarreling over territories that they will guard jealously until winter hardship brings them all into the yard again. Our feeding has produced a heavy overpopulation of squirrels on

ah

yes

our small acreage. If we should desert them in winter, many would starve. When we leave this beautiful place, and some day we must as the years pile up, we will go in summer, when the days are warm and wild food plentiful. Then our squirrels will have time to adjust to the disappearance of corn and crackers and to spread out and set up housekeeping on territories large enough to support them.

The chipmunks are so amply supplied that they retire to their winter dreaming at the first cold snap. In the space where the carrots have been taken up there is a fresh growth of wild plants, and ruffed grouse and spruce grouse pick the unexpected treat in the midst of the surrounding dried vegetation. As our winter birds gather, we check to be sure that we have enough food for all visitors during the coming months.

note this

This is very important, not only because of the responsibility we have toward the squirrels, but because birds that become accustomed to feeding at a location will depend on that source of food. If the food does not appear, the birds wait, often so long that they are too weakened by hunger to find another supply. Birds can survive bitter weather if they have enough food to keep their high-speed engines running. They starve before they freeze. Anyone, anywhere, who feeds birds in winter, should put out something for them *every day*.

I do

In the year beginning October, 1959, we used eight hundred pounds of cracked corn, sixty-five pounds of suet, seventy-five pounds of graham crackers, and twenty-one pounds of ground beef, along with assorted leftovers like macaroni and boiled potatoes (moist foods like these freeze solid quickly in bitter cold weather, so should be offered only when there are customers to eat at once), pancakes, bread, cookies, fudge crumbs (a special joy to the flying squirrels), bacon rind and grease, other meat and fat scraps, and turkey

1959

gravy (which the squirrels love). True, it makes a deep dent in the budget—but nothing good comes without its price. In that year we had ninety-one birds and eighty mammals as regular guests, and more than two hundred birds and ten mammals as occasional visitors. This does not count the flocks of siskins and grosbeaks that passed by, nor the scurrying shrews, nor the mice and voles. Some of these wild things brought us laughter. All of them brought us knowledge. They have repaid us a thousand times by teaching us gentleness and compassion.

ch, yes

Some Year-Round Mammals

A BABY RED SQUIRREL dropped past my nose as I opened the door. He had crawled from the nest under the dormer and fallen eleven feet, to thud, spread-eagled, on the stone step. Faint rustlings from above told of others in the nest.

While Ade went after a board to nail above the door as a sort of porch to catch any more adventurous youngsters, I took the unconscious little fellow inside. He was well-furred but still very immature and, as I examined him, he revived and began to quiver with the pitiable terror that comes from exposure to strangeness. I held him gently against my breast with one hand, covering him lightly with the other. Soon the warmth and darkness and protective enclosure had their effect. The quivering stopped and the wild racing of his heart steadied to an even beat. I put him on the counter, where he sniffed out a small piece of graham cracker and made a strong assault on it with his partly emerged teeth. Apparently mother squirrel was weaning him on this delicacy.

Ade had finished his impromptu carpentry, which included a slanting board to serve as a ramp by which the lost child could get back home, when the mother dropped from the roof onto the feeding shelf by the door. She was unusually handsome, her upperparts clear red-brown, her underparts spotless cream-color, and her black side line clean-edged. She nibbled a cracker as she watched the proceedings, her only sign of excitement a twitching of her flaring tail. When I held her baby out to her, she sniffed him competently, then observed as I set him on the ramp and he crawled back through the nest opening. After an inspection of her household, she scampered away. Ade and I waited to see what she would do about this intrusion into her nest that was obviously no longer safe.

Soon she returned and pushed a baby onto the ramp. She spun it round and round until she got a firm grip on the belly with her teeth. The youngster curled its feet and tail around her neck like a furpiece, and she climbed down the wall and ran laboriously thirty feet to a low wooden shelter over an unused engine, quite a task for a mother only a foot long, including her five-inch tail, and weighing about six ounces. She moved her three babies in ten minutes and inspected the nest to make sure that she had left none behind. After all, squirrels cannot count. Then she rested, grooming her tail atop the engine shelter, which resounded to the play of her children. (We wonder whether she may have selected a ground-level site for her new home because she was now aware of the danger from height in her old one.)

When a neighbor learned that squirrels nested in our roof he told us that he had seen a cabin where they had entered through a cardboard-covered broken window and created havoc—oil lamps upset, curtains torn, toilet paper shredded and draped around with the abandon of serpentine at Mardi Gras. His own seedling maples had died after squirrels chewed the bark to lick the sweet sap in spring. It is obvious that if the windowpane had been replaced or the opening covered with metal, and the trunks of the young trees protected with screening until the maples were established and strong, no damage would have occurred. One cannot expect wild things to have consideration for human property rights.

We were also told that, if we did not get rid of our squirrels, we would not have any birds. No doubt squirrels eat some eggs, but that they remove any native bird species is refuted by the large population of both that live on our property and gather at the feeding station.

yes

Not long ago I read of a man who found a red and a gray squirrel fighting at a nesthole in his garden. Because of the red squirrel's reputation as a nest robber, he shot the red. The gray ran away. When he looked into the nest, he found a litter of baby *red* squirrels, one killed and another bitten. A neighbor played foster mother to the living babies but, had the hasty shooter considered that the gray might be the aggressor, or, better still, had he investigated instead of shooting, the small mother could have reared her young and the man would have had the pleasure of watching the whole family in his garden—and red squirrels, with their bursting energy, sputtery tempers, and individuality, are fine company.

The Butler waits quietly by our step until less mannerly squirrels have stopped squabbling over crackers and departed. Then he stands up, places one paw across his white vest, and bows sedately from the waist. The effect is ruined on those occasions when he bows so deeply that he falls on his face. The Outfielder jumps up and down on stiff hind legs, smacking his forepaws together so that you can almost hear him yell, "I've got it!" The Little Old Lady has reared four litters in our yard. For three years she held an advantageous nesting spot near our food supply but lost it in fair fight last spring. One of her ears is torn, the tip of her nose has disappeared, and her tail is bobbed. When she sees one of us at the door, she hurries up, limping a little and stopping now and then to wave a forepaw, as though to say, "Wait for me!" We give her a bounteous share because her slowness will be her undoing one of these days.

Red squirrels vary as much in appearance as in behavior. They may be any shade from tannish-gray to brownish-red, and some of ours, possibly due to inbreeding, have lost the black line, along the side between the dark upperparts and the pale belly, that is a distinguishing characteristic of the species in summer pelage. In winter, the fur thickens, the black line disappears, and the eartips bear tufts of hair. The tails may be anything from thinly furred dull brown to wide, bushy red. The notion that those with thick, bright tails are fox squirrels is wrong. The fox squirrel is a two-foot-long mammal of the east and south, whose coat varies from red to gray and partly black, depending on its location.

Our red squirrel colony produced a lovely, snow-white albino several years ago that was received with suspicion by its normally colored fellows. It was live-trapped by a summer neighbor and presented to a zoo, where its peculiarity was an asset instead of a handicap. The gray squirrels, familiar beggars of city parks, are found in Minnesota but are not native to our location, probably because there are few large nuts and seeds on which they might feed. The "black squirrel" is a melanistic form of the gray, although black individuals may occur in other species.

As the red squirrels jump from branch to branch overhead, I wonder how many deer have been saved from starvation by the cedar cuttings that fall as a side-issue of the little mammals' search for cones. As I see one of these squirrels carefully burying a choice bit of food, I wonder how many trees have grown from seeds that were lightly covered by the patting of their small forepaws and then forgotten. As I observe the serious business of removing burrs from one's fur, I wonder how many flowers have bloomed in new places from seeds carried behind fuzzy ears. And, when someone says red squirrels are too common to be interesting, I wonder how many people have missed their touching and amusing antics simply because they carry no aura of the strange.

nice thoughts here

After the red squirrels have gone to sleep in their solitary nests and darkness settles over the forest, the northern flying squirrels wake in their hollow trees, where large groups live companionably together. One by one they poise at the nest opening, often an old woodpecker hole, then leap into the air. Their flight is really expert gliding, controlled by varying the tension of the furry membrane that extends on each side from the fore to hind leg. Sometimes we feel the soft wind of their passage, and we have seen them gliding almost horizontally on short leaps from tree to tree.

Shy and nocturnal, the color of the night, they are rarely seen alive. This is fortunate for them because they are so delicate that they can be fatally injured by handling, and so sensitive that they may die of shock when captured. When we hear the soft thumps of their landings on the feeding shelf, we are careful not to frighten them, and we do not throw blinding light into the woodshed where we hear their faint *chp-chp-chp* sounds. Because their "wings" prevent them from swimming, we never leave uncovered vessels in the open. They can drown in a pail of water, as can chipmunks, mice, other small mammals, and insects.

The flying squirrel's fur is elegantly thick and soft, creamy on the underparts and olive-brown above, with a dark outline along the outer edges of the wings. The tail looks like a long and densely furred flat plume and makes up a little more than half of the ten- to eleven-and-a-half-inch overall length. (Southern flying squirrels are lighter in weight and have an inch less tail.) Their eyes are like small, black moons with dark shadings above and below, faintly like the triangular marks that are a part of a clown's make-up. Their expression is appealing and winsome. As they eat, they sit hunched forward, their tails upright with the tip bent back. From the front, where the folds of their "wings" add to their width, they resemble chubby, contemplative Buddhas.

Flying squirrel mothers are extremely devoted to their

young. There are usually three or four, and they are born blind, deaf, and naked, but with full "wing" membranes. They wear baby fur at two weeks, see at four, and are weaned at five. At three months, they have their adult coats. Apparently they are not brought to our feeding shelf before this age, for all the young we have seen look like small adults.

Sometimes a half-dozen of these squirrels crowd peacefully together on the feeding shelf. If I open the door, heads appear out of the tangle of fur and tails. The first squirrel to extricate itself clings to the shelf with hind feet and clutches at the doorpost with brown-gloved fingers, sometimes losing its grip and hanging head downward while it reaches out to my hands and the graham cracker I am offering. One by one, they take their crackers and spring away. I hear soft crunching as they eat, two of them on the woodshed roof, another on a claw-hammer hung just beneath, one on a log-end at the cabin corner, one behind the split stove wood, and the last inside an overshoe on the shed floor.

Although they push and shove to get at the crackers, often climbing over one another's backs, I have seen only one fight. Two squirrels waited but I had only one piece of cracker, and I gave it to the nearest squirrel. As it turned to "fly," the second squirrel caught the opposite edge of the cracker in its teeth and a see-sawing tug-of-war resulted. The squirrel that had the prior claim suddenly drew back a forepaw and boxed the other's ear smartly. As though amazed at such unprecedented behavior, both drew back, hesitated, then pounced in a flurry of claws. I laid my hands gently on their backs, to touch fur whose softness I shall always remember. They froze, then turned to look cautiously up at my huge hands. Like chastened children, they politely accepted crackers that I offered to both simultaneously, just in case there might have been a resumption of the fray.

The almost unbelievable quickness of these creatures is in sharp contrast to their quiet, gentle manner. Sometimes they replace one another on the shelf so rapidly that I cannot see the exchange. I am aware that the squirrel I have given a cracker to is gone because the squirrel now sitting in the same place has no cracker and is of a slightly different size or appearance. Now and then an individual eats corn on the ground and, as I watch, seems to vanish. I follow its passage toward me by the dipping of the cedar branches. If it is one of my luckier nights, the squirrel drops to my shoulder to take its bit of cracker. Then I am made aware of its going only by its absence and the pressure of its propulsive spring.

The flying squirrels do not leave their nestholes before dark unless disturbed and, even with this protection and their great speed of movement, they remain cautious. If they do not come to the shelf by midnight, I look for the tracks of a hunting fisher; if there are no fisher tracks, I listen for the hoot of a barred owl. Now and then we find flying squirrel tails lying forlornly on a path, but this is the way it was meant to be. We feed both the hunters and the hunted, confident that we will create no irreparable unbalance so long as we do not play favorites.

Sometimes Ade and I hear a brisk pattering in the double roof, which may be followed by a cushioned plop as a mouse drops from the chimney bricks into the woodbox. Less often we glimpse a small figure streaking through the shadows. Very rarely, because deer mice are as timid as they are small and dainty, we have a tame mouse.

Little Brownie arrived on an evening before Christmas when our coffee table was covered with candy, cookies, and nuts mailed to us by faraway friends. When we heard the faint rustling near our feet, we sat motionless. Very slowly, looking from one of us to the other, the mouse climbed to my chair arm, crouched two inches from my hand, and jumped to the laden table. She nibbled the edge of a cookie, licked a small portion of filling from between its layers, and ate a crumb of milk chocolate. She jumped to the rug and was next heard rattling discarded nut shells in the wastebasket. We did not hear her go.

The next night we placed three hazelnut shells on the table, filled with bits of cookie, chocolate, and walnut kernel. Little Brownie arrived at about the same time as the night before, and continued to come to supper for five weeks. She chose her meal, sometimes leaving the cookie or chocolate but always eating the nut. She soon lost much of her timidity and groomed herself on the table after eating.

She washed her face with her forepaws and cleaned each whisker thoroughly, then went over her brown coat and soft, white vest. She cleaned her tail, which was well furred and sharply bicolor with dark top and light underside, then pulled it around for a careful inspection. She inspected her spotless white gloves and stockings and, as a final touch, smoothed the backs of her large, translucent ears upward, in a gesture reminiscent of a lady giving that last pat to her hairdo. And all the time she watched us, her big, black eyes bright and sparkling as the waters of a stream running dark between snowy banks.

These mice love corn and we often see them feeding outside at night, their eyes rose-colored in the lantern light. Sometimes they move about by day, and we may see their tracks on the snow overtaken by those of a weasel or mink. By spring, when their prolific breeding starts, their numbers are reduced to a level within the limits of the seed crop that furnishes most of their food. They sometimes mate at only five weeks, and could produce six or seven broods of from two to nine young during the northern May-through-October breeding season. Actual numbers vary, depending on weather, food, disease, and the incidence of carnivores.

The few mice that enter the house cause no trouble. They will nibble new shoots on some house plants, and have a fondness for chenille, cotton yarn, and toilet tissue for nests, but these things can be kept out of their way. They do not chew indiscriminately as house mice do.

The terms "deer" and "white-footed" are sometimes used interchangeably. Deer mice, which may have brown, tan, or gray upperparts and are three to four inches long from nose to rump, with a two- to five-inch tail, differ superficially from white-footed mice only in that the white-footed have shorter, less strongly bicolored tails.

There are two species of voles here: the meadow, often called meadow mouse, and the boreal red-backed, also known as Gapper's red-backed mouse. They are from five to seven and a half inches long, of which one and a half to two and a half inches are tail, but they look shorter because they sit hunched into a ball. They resemble fat, bobtailed mice, with small eyes, and ears buried in long hairs. Color easily distinguishes the tawny phase of the red-backed boreal vole from the gray meadow vole. However, there is a gray phase of the boreal species in the north and east. The meadow vole reaches a longer maximum size, but the size range of the slightly smaller boreal vole is encompassed by that of the meadow vole. Both are neat, clean, little creatures, storing grass stems for food in piles along their runways and leaving their droppings in places set aside for them.

Ade and I are particular about food storage—everything is kept in cans or jars or tight-closing wooden cupboards. True, mice can gnaw through wood, but their teeth are so small and their jaws so weak that it would take generations of them to get through a half inch of hard board. Our only invasion came when we were suddenly called away. In my hasty preparations for departure, I left a box of grahams on the kitchen counter. Ten days later we found the package empty and the counter cluttered with droppings. But even this was interesting because, among the black scats of the mice and the brown ones of the voles, there were little *green* scats, which told me that bog lemmings lived nearby.

The lemmings are superficially similar to meadow voles, except that they are sometimes a bit larger and have shorter tails, usually less than an inch long. The southern bog lemmings are most common here and the northern appear occasionally, but these species can be separated only by a careful examination in the hand. The bog lemmings are cousins of the brown and collared lemmings that inhabit the far north, principally the tundra. The collared lemmings are the only small rodents that assume a white winter pelage. Lemmings are a staple food of the snowy owls, and decreased lemming populations in the tundra may cause the migration of these birds into our northern states in a so-called "snowy owl year."

Northern Minnesota has none of the harmless, interesting wood rats, but Norway rats scavenge in the towns along Lake Superior. If these rats should be transported into the woods, unsightly and unhealthful dumps where some resorts and residents dispose of garbage and trash will supply excellent food and lodging. Ade and I destroy all our combustible trash and burn out our cans in a sheet-iron stove, which also helps with our winter heating. Ade then flattens and buries the cans, along with anything else that will not burn. Food scraps are fed to our animal family. All this takes time and labor, but we do not wish to contribute to the defilement of the forest.

All true rodents have two gnawing incisors in the front of the upper jaw, but rabbits, hares, and pikas have four, and this sets them apart in the special order of lagomorphs. Our representative lagomorph is the snowshoe hare, sometimes called—along with others of his kind that change to white fur in winter—the varying hare.

During the summer, a brown head with four-inch, black-tipped ears, and eyes like ripe olives, lifts above the ferns to make a survey of the surroundings. Then the hare comes into view, a brown animal some twenty inches long, with huge brown feet and an insignificant brown tail. As he rears on his haunches, showing some white on his belly, his short front legs look ridiculously small. Always watching, always careful, he turns his round head as though on a pivot. Reassured, he hops leisurely away, his hind legs unfolding awkwardly and the underside of his tail showing a fleck of white. He eats a meal that consists of five of my largest violet plants and thirty-eight oversize dandelion leaves. Each leaf is nipped off individually at the base, then nibbled slowly into his mouth as his nose wobbles in time with the sideways movements of his chewing. The silence is disrupted by a loud "YAHOOOO!" from some exhilarated canoeist near the shore. The hare kicks up and out with his clawed hind feet, in case the danger might be near. He enters the forest with enormous leaps, all traces of awkwardness gone as his powerful muscles go into real action.

His long, strong, hind legs give him the speed that is his best means of defense. They carry him to some well-known refuge—under a tree root or brush pile, perhaps. There he sits motionless, ears laid flat, almost invisible against the forest floor.

Once a very young hare, not long away from the fur-lined form that had held it and its several brothers and sisters, was hopping outside the door as Ade and I walked down the path from the road. Terrified, it did the best it could according to its instincts and short experience. It hopped into a corner formed by two walls and froze on a flat, white stone. Its brown fur was completely conspicuous and there was no cover except the shadow of one wall. We walked past and inside as slowly and quietly as we could. For three hours the baby hare showed no movement except that of its eyes. When dusk came, it crept across the open space and under a thick growth of jewelweed. Had we really been enemies, this helpless creature's mistake would have been its last.

When summer is waning, a hare will come down a path on white feet. Over a period of eight or ten weeks it will continue to change from brown to white in a patchy manner. This mottled coat is very successful camouflage against the partially snow-skimmed ground of spring and fall. I once saw a hare so nearly invisible against the splotchy April snow that, at fifteen feet, my sight was attracted to it only by one shiny, dark eye. The hare's white coat is dangerously prominent if it completes its change before the coming of the snow, but against the later shadowy drifts the animal looks like a carving wrought from the snow itself. The black eartips serve, as does the ermine's black tailtip, to draw a pursuer's attention away from the body.

No night is too cold for the foraging snowshoe hare. Sometimes, when the thermometer shows forty below, a moon shadow draws my eyes to one as it is eating rich, warming corn. Hare tracks are in the clearings every morning, the marks of large feet and of small, a companionable symbol of our northern winter.

If our carrot patch had not been so attractive to hares that fencing was a must, Ade would not have unearthed the star-nosed mole as he was preparing the ground for the posts. He brought the six-inch, shrewlike creature inside so that we could examine it closely, for these energetic tunnelers spend all their lives underground and in the snow. This apparently helps them to survive because, unlike most small mammals, they have only one litter of from one to five young in a year; the babies are born in a deep tunnel in early spring.

The mole's gray-black fur was so soft that my fingertips registered no sensation when I touched it lightly. Its tail was flattened, well-haired, and shaped like a slender, pointed feather. Its heavily clawed broad front paws were not turned outward as is the case with many moles. Its eyes were the veriest pinpoints and I could not see any ears. Its marvelous feature was its snout, rimmed with eleven pairs of symmetrically placed flesh-colored tentacles, so that it looked like a flower or a sea anemone. The tentacles were of different lengths and shapes, probably to adapt them to different uses.

The mole paid little attention to my stroking hand but when Ade reached out to touch it, it became very excited, nuzzling his earth-stained fingers with its remarkable nose. He laughed, went out, and returned with the great-granddaddy of all earthworms. He placed this on the counter behind our guest, which sniffed, turned slowly, and pounced. Holding the struggling worm firmly, the mole used its digging claws to scrape it clean of dirt, and devoured it to the last segment. Then, thrusting its tentacles into every cranny and upsetting the pepper, fortunately over its tail instead of its nose, it made a tour of the counter, presumably hunting for more worms.

We took it back to the garden, where its powerful forepaws took it out of sight in a shower of soft earth, to go about its business of digging worms and eating a bit of root now and then as a vegetable. All year round it would be tunneling, ventilating and loosening the sparse forest soil, adding its bit of cultivation to the benefit of future vegetation.

The porcupine is so large and bristly and moves with such awkwardness that it is hard to think of it as a rodent, a member of the same order as the sleek and agile squirrels and mice and voles. One I saw in the yard in 1957 may have been the last in our area, as porky is rare here and I have heard of none since.

Frosty black rear wagging, it waddled along as though it were completely fagged out. When I hurried after it, its quills puffed up until the animal looked like a worn-out brush. It flipped its tail in warning and bared its powerful incisors. It was about two and a half feet long and may have weighed twenty pounds. It looked frightening, but was plainly so unhappy in my presence that I got it a piece of salted bread and returned to the cabin. Porcupine and bread were gone a half-hour later. A month later, a freshly cut pine twig dropped before me in the path. I looked up and saw the porky, a black ball high in the tree, but there was no sign of it when winter came. I regretted its going. Its cuttings from the evergreen would have done much to help deer and hares through the winter, and I should have enjoyed getting acquainted with it—quills, crotchety manner, and all.

We had nothing to fear from porky. Porcupines do *not* throw their quills, which come out only on contact. Porky could not chew up our ax handles or paddles or other sweat-salted wooden equipment because Ade does not leave things lying around outside. I do not fling dishwater out the door, letting dribbles make the sill attractive as porcupine food.

There is some question in my mind as to the complete cause of the porcupine's rarity here. Fishers certainly help to control its numbers, for it is one of their favorite foods. When a fisher catches a porcupine, it turns its prickly dinner onto its back and eats from the unprotected belly side. Even so, quills penetrate the fisher's outer skin. Many of them pass out again at the surface without injuring the inner skin or muscles. Some pass harmlessly through the digestive tract. If numerous quills penetrate the fisher's mouth and throat it will probably die because it cannot feed. However, we saw fisher tracks near here for the first time in January, 1956, some years after the porcupine became scarce. Perhaps disease is a factor. In any case, Minnesota game wardens are live-trapping fishers, which the U.S.

Forest Service then sends to locations where the animals had been exterminated and where porcupines have increased to such numbers that they menace timber stands. Similar relocations are being made elsewhere by other conservation agencies.

This move has great significance. The value of the predator in forest balance is at last being recognized. This may lessen the persecution of foxes, wolves, coyotes, and other meat-eating mammals. These carnivores, if left alone, will control the herbivores that, in over-large numbers, can seriously damage the vegetation of their environments by eating both plants and seeds. Eventually, as the vegetarian populations are reduced, the carnivores will decrease in numbers from lack of food, and a natural, self-maintaining balance will be established, the kind of balance that produced the wild richness that early settlers encountered.

If this comes about, porky's tree-barking will have played an important part. Perhaps, as porcupines decrease and deer suffer from the absence of their cuttings, men will gain an understanding of porky's value and, from this, recognize the value to the forest of all its inhabitants.

The Deer Family

BEFORE the white man's coming, vast stands of pine swept inland from Lake Superior's North Shore and supported four members of the deer family: elk, moose, woodland caribou, and, in the more open southerly inland areas, white-tailed deer. Permanent settlements frayed the edge of the forest during the last half of the nineteenth century; lumbering began in earnest around 1890. Moose and caribou were then common. Whitetailed deer were rarely encountered, either by settlers along the shore or by lumberjacks, trappers, and prospectors who followed the lake chains and animal trails inland. By 1920, when Paul Bunyan's axmen had reduced the forest to token remnants and fire had added to the devastation, great changes had taken place in the distribution of the four types of deer.

The elk, probably never found except in scattered small herds, were gone. A herd was introduced later near Lake of the Woods and a few of these may still lift their wide and graceful antlers and show their conspicuous white rump patches in secluded parts of that area.

As fire destroyed Minnesota's foliose lichens, the caribou moved north. The last one reported in the state was seen in

1921 at Iron Lake on the Canadian border. Now, the caribou nearest our home are in Ontario, a little more than a hundred miles away, where they are increasing in number. As the forest areas that are preserved by the Federal government in their primitive condition increase in age and tree size, reindeer moss may again carpet their earth. There is some growing near our cabins. Perhaps some day the caribou will trot south across the ice into the land that once was his.

The woodland and barren-ground caribou bulls have antlers unlike those of any other deer. Two heavy beams reach out and up from the head and turn forward at their upper ends. These ends bear points, or tines, up to a foot long. The tines often are branched and may be somewhat flattened and widened, or palmated. A pair of tines, one or both of which may be vertically palmated, extend forward from the bases of the main beams over the nose. No other deer has forward-pointing tines. The caribou doe is the only female American deer to bear antlers, though, rarely, does of other species have abnormal antlers that do not fully develop. Even the normal antlers of the caribou doe are smaller and less elaborate than the bull's.

The woodland caribou, formerly found here, is brown, with whitish mane, some white on the rump below a stubby white tail, and white just above the rounded, outsize hoofs. It bears small resemblance to the other deer of the area. The barren-ground species, found in the Arctic, has smaller antlers and a whitish coat.

The upstanding and leafy lichens that grow both in the Arctic and on the floor of mature northern forests are the caribou's principal food. Their big hoofs dig through the snow to the "moss," and the antler projection from the brow, although it does not extend far enough forward to sweep a path for a foraging nose, may help keep the way clear in

deep snow. Thus the Indian name "caribou," from a word meaning "one who shovels."

Moose belong to the coniferous forest and, as alder, birch, and maple flourished on cutover land, they almost vanished in Minnesota. In recent years, however, balsam fir and spruce have covered large areas and moose, under protection, are increasing in spite of a nervous disease, poaching, and the casual shooting that offers some dark and obscure satisfaction to those who enjoy killing for its own sake.

There was much excitement on our side road in the summer of 1955 when a moose track was seen. The next winter, as I stepped outside the cabin, I saw my first moose standing in lantern light not ten feet away. He still had his antlers, almost as wide as I am tall, and he peered down at me over a nose so long and so high above me that it reminded me of a stub branch on a tall tree. His size so startled and overwhelmed me that I could not move. It was not until he snorted in a faintly contemptuous manner and trotted off that I realized that a snowpile he stood on had added three feet to his height. In the morning I measured his tracks, clear and sharp in the packed snow. The paired teardrop marks were nine inches long instead of the average seven, and they were spaced six feet apart. He was a veritable king of bulls.

The exclusively North American moose are the largest deer on earth, so they do not have to be giants of their kind to be spectacular. The bull's heavy body rises above his thin, knobby, four-and-a-half-foot legs to six, and even seven, feet at his shoulder hump, and a species found in Alaska may be eight feet at the hump and weigh almost a ton. When a bull moose lifts his enormous head, with its wide, flat, palmated antlers and his bell, a two-foot flap of skin and hair that hangs beneath his throat, he commands respect, even against a background of hundred-foot trees. In winter, after he has shed his antlers, he is still imposing and, in early fall, when his thickening winter coat adds to his bulk and his newly grown antlers are still covered and thickly padded with velvet that has a greenish tinge, he is almost unbelievable.

The only time when moose are not unmistakable is when they are eating water plants in shallow water. Then, as they stand body-deep with heads ducked under to feed, they look like large, dark rocks. There are moose in the swamp south of our cabin. Tracks appear on the road at irregular intervals and occasionally we hear a musical *blaaaaat,* like a sour note on a cornet. Only once have I seen a moose in the yard in daylight.

She stood on the path, some twenty feet from my window, a smaller edition of the bull, without antlers and with a shorter bell. Her head was lifted and, with much wiggling of her overhanging upper lip, she sniffed the air. She was watching Ade launch the boat, much as a whitetailed doe might, moving her ears to isolate and locate sounds. He did not see her and started to walk in her direction. She drew back her head, whirled, and trotted up the path. Any notion that a moose is awkward disappeared as I watched. Her neat, dark-brown rump, with its tiny flat tail, was held almost motionless, while her big feet shuffled along with the speed and grace of a soft-shoe dancer.

She turned aside into the tall growth of maple and alder with the characteristic soundlessness that so contrasts with the moose's great size that it is uncanny. She bent her head to nip off a leafy branch, moved on a few steps, and simply disappeared. The brush was no thicker and the shadows no darker, but she was no longer there.

When the fall rutting season arrives, the bull moose is neither silent nor amiable. His mating bellow sounds like a diesel horn and, I think, may carry as far. The first time Ade

and I heard it we almost dropped our teacups at the blast of
sound that seemed to originate just outside the cabin. After
a reasonable interval of quiet, we investigated. Some five
hundred feet away we found moose tracks, leading into thick
growth from a road surface torn by pawing hoofs. From beyond a ridge, we heard a confusion of grunts and bellows.

We climbed. The ridgetop overlooked a small valley covered with grass, brush, and clumps of young aspen and
birch. Two bull moose were fighting—panting, grunting,
pushing furiously antler to antler. Suddenly one was shoved
backward. With a scream of rage, he landed heavily on his
rump. Before the other could take advantage of the fall and
get around to his side where he might strike a damaging
blow to the ribs or belly, the first was up and the two squared
away. With lowered heads they appraised each other like
boxers or wrestlers seeking an opening. Abruptly the great
antlers crashed together and the fight was on, accompanied
by terrible roars and an ever-present humming sound. (A
trapper has since told me of a similar contest he witnessed,
where the humming was so loud that he had the sensation
that invisible hornets were buzzing around his head.) They
tore up the ground and snapped off trees, for the battle arena
moved to a new location when one or the other ran a hundred yards or so, seemingly to catch his breath before renew-
ing the struggle.

The bulls were well-matched in size, with fine antlers,
behind which their bristling neck hairs added to their angry
appearance. The wind changed and the bulls scented us.
They stopped fighting and trotted hurriedly away, no doubt
to resume their battle in more favorable surroundings. Probably one contestant eventually gave up and ran away. Rarely,
one moose may kill the other or they may lock antlers and
die together of starvation. But we could be sure that somewhere nearby a cow was awaiting the outcome of the battle.

Over and over one hears and reads that there is nothing

in the North Woods that will hurt you—I am ignoring the sensational type of reporting that turns a bobcat into a Bengal tiger. It is also said that the only thing you have to fear is yourself. As with most platitudes, there is some truth here, in that no one would be frightened or injured in the woods if he did not put himself in a vulnerable position. This sort of thing glosses over the important fact that the woods holds things to be avoided. The bull moose, during his fall rut, is one of them.

Moose behave in individual ways, according to temperament. Some are retiring, even when driven by the mating urge, but others seem to look on any sizable, strange, or moving object as a rival. In November, 1961, a friend of ours, driving his two-ton truck, saw a bull walking along the roadside, accompanied by the cow that was his conquest of the day. Instead of driving on, our friend stopped and blew his horn. The bull answered this challenge with a mighty bellow, pushed the cow roughly aside, and charged. Happily, the truck's engine was idling. As our friend got under way with all haste, the bull's massive head almost crashed into the truck. Moral: Do not contest the right of way with a thousand pounds of bone and muscle.

Had Ade and I moved to this forest in 1854 instead of 1954, we might never have seen a whitetailed deer outside our cabin door. These beautiful mammals, charmingly painted Christmas-card scenes to the contrary, do not usually inhabit mature coniferous forests. Before 1870, they are said to have been absent from the North Shore of Lake Superior. Only in the latter years of the nineteenth century, as a growth of suitable low browse covered the cleared and burned land, did the whitetails move northward along the shore and migrate inland. The changes that take place in their numbers as the carrying power of the land waxes and wanes is very clear in the area around our home.

The land stretching west from a half mile beyond our property was burned over fifty years ago. Fireweed and horsetails, grasses and ferns, daisies and everlasting, dandelions and violets, started revegetating the burn. Raspberries shot up, and brush followed: maple, willow, elder, mountain ash, dogwood, and sumac. Then came birch and aspen sprouts, cedar seedlings starting their relatively slow growth, furry little spruces, and balsam firs.

Deer browse is plentiful in summer but, when cold and snow limit the supply, high nourishment foods—white cedar, mountain maple, mountain ash and red osier dogwood— *winter* are necessary. On the burned-over land, the regrowth of good winter browse reached its peak in the thirties and early forties. The deer in the area, their numbers controlled by wolves, walked their trails through the snow, nipping off branch tips as they went, pruning but not destroying the vegetation, and chewing their cuds in well-fed contentment.

The whitetails multiplied and more and more people entered the area. The wolves retreated except during the man-scarce winter months. The trees grew taller. In 1950 there was a severe winter, followed by a long, cold, snowy spring. Overbrowsing occurred. Many of the deciduous shrubs were destroyed and young pines were damaged. The hungry deer even ate balsam fir; although a nutritious food for moose, balsam is only starvation stuffing for deer. Larger deer were able to reach the lower branches of the cedars by standing precariously on their hind legs, but smaller animals, and especially fawns, could not. At accessible, flat places along the shore there are lines of cedars, neatly trimmed to a height of seven feet. This is not the "clearing up" of summer residents, but the browse line left by the deer as they overtaxed the ability of the land to feed them. A starve-out was inevitable and numbers of the deer perished. A saddened ranger told me of finding twin fawns frozen as they cuddled together in their bed in the snow simply because they had not been able to find sufficient food to maintain body heat throughout a bitter night. From this time on, as the trees grew larger, fewer deer have inhabited the area.

To the east of the burned-over land is mature forest. In dense stands of pine, balsam, and spruce the ground is almost free of small growth. The cedars are huge, and bare-trunked for many feet up. The birches are almost a hundred feet tall and most of the aspens have rotted, been used for a time as woodpecker trees, and fallen. It has been years since this old forest had food for many deer.

Now, where the fallen trees let light to the ground, maple and mountain ash are springing up, and browse is growing again in the clearings near summer homes. Much of this is cut and burned by people who want increased ventilation, but Ade and I let our brush grow except adjacent to our cabins. This, along with the corn we always have waiting outside, helps maintain a few deer and gives us the opportunity of watching the remnants of this shore's herd.

In 1956, small groups of eight to twelve browsed through the area all winter. In 1957, they passed by twos and threes. In 1958 and 1959, we had six tame deer in the yard, and Ade noticed tracks of others crossing the road about three miles west in the burned-over area. In 1960 the tame herd was still here but the tracks to the west were absent. In the fall of 1961, hunters reduced the tame herd to three, our old doe, Mama, and her twin fawns. During the following winter, Ade saw few tracks on the side road. A single deer can leave many tracks, but if the tracks are few, it is certain that the deer are, also.

When the deer were numerous, one of the first signs of winter was the eerie and beautiful howling of timber wolves. Ade saw their tracks along the road when he was carrying our mail, and heard ravens croaking as they finished off wolf-killed deer carcasses. Since 1959, we have heard only very distant wolf songs and only one wolf has been seen nearby.

As long as the food supply is adequate, a deer herd maintains itself in the presence of normal predation by wolves and controlled predation by men. It is common to blame wolves for the disappearance of the deer, but only when the browse becomes inadequate does predation reduce the deer's numbers seriously. Even then, the wolves do not take the last of the herd, possibly because the deer are so widely scattered that hunting them is too difficult, possibly because instinct impels them to withdraw in protection of their food supply.

Mama, our tame doe, is so named because she is a careful and conscientious leader and teacher of her fawns. The pair born in June, 1960, were bucks. The one with the vast appetite and few manners was Pig. The shy one was Brother. On New Year's morning, while Ade hung the suet feeders, the deer trio watched him from a bank. He put out fresh corn, and Pig bounced down to feed. Brother walked down and nosed into the corn pile where Pig was guzzling. Pig reared and struck at Brother's face and just behind his right front leg. Mama came off the bank with a bound that cleared the whole open part of the yard. Pig, the very picture of guilt, started to leap aside but, when Mama turned to Brother, who had limped away, Pig ducked his nose into the corn again.

Mama touched her nose affectionately to Brother's, then licked the bump on his cheek and nuzzled his side, as though to see whether he were injured. She led him to another corn pile, walked over to Pig, and butted him away from the food. Startled, he reared up angrily. Mama stared at him with as near an outraged frown as a deer may manage and thumped him briskly on top of the head with her hoof. It was a good thump—loud and solid. He shook his head, rubbed the bump with a hind foot, and tentatively nosed toward the corn. Mama's hoof was lifted and ready. Pig retreated to the bank, while Mama and Brother finished breakfast.

When they followed their snow trail back to the bank, Mama brushed past Pig as though he were invisible. He approached and she turned away. He stood alone, his head drooping in utter dejection. Now and then he looked toward Mama, who did not respond, although Ade and I, from the window, could see her eyelashes move as she glanced at him. Some twenty minutes later she moved in his direction as though by accident. He glanced up, then hung his head again. She turned directly toward him and raised her head from the fallen cedar leaves she was eating. Pig moved a small step nearer, lifted his head in stages, stretched his neck toward her, and received an affectionate lick on the nose.

Pig jumped back to the corn, with Brother close behind and Mama keeping a watchful eye on them. When Brother approached, Pig moved politely to another corn pile, and Mama regarded them with a gentle, complacent expression, as though she felt she had done well.

Then she stiffened, turned her ears toward the road, and thumped with her front feet. The fawns sprang to her side, tense and listening. A car stopped at our path. Mama bounded away, with Pig and Brother at her heels, white flags waving.

Does may mate at any age from one and a half years on and usually bear twins, born in May or June. Single fawns sometimes come from first matings.

The rutting season extends through October to a peak in

November. Then the bucks, who have summered away from
the does and fawns, come out of seclusion, their antlers fully
grown and their necks thickened and strengthened for fighting their rivals.

The antlers appear as knobs in March and grow through
the summer, covered and nourished by the fuzzy skin called
velvet. In late August, the buck removes the velvet from his
new weapons by rubbing them against a tree. The number
of points does not indicate the buck's age, and first antlers
are not always single spikes. One pair of Mama's fawns
wore six- and seven-point antlers when they were a year and
a half old. Food quality affects antler growth, and our feeding may have contributed to the young bucks' fine heads.
The male sex hormones and hereditary factors control and
direct the growth. All of Mama's fawns have shown certain
markings and other resemblances to her and, if the yearling
bucks' antlers were a reflection of their father's, he must
have been a very handsome fellow.

Bucks are completely polygamous, and collect neither
harems nor single mates. Although they may winter with
the does, and even help a mother train her fawns, there is
not necessarily any connection between such companion-
ship and the mating that produced the maturing fawns or
the one that will produce new fawns next spring.

Mama and her twins born in 1958 learned to feed with
us in January of 1959. She departed in May, and returned
with her new five-month-old fawns in November. On June
10, 1960, I started to walk into the woods a hundred feet
from the log cabin and was confronted by Mama, forelegs
spread, head lowered, snorting and ready to chase me home,
which a doe is perfectly capable of doing. Then I noticed

that the heavy roundness of her pregnancy had disappeared since the previous day. Mama was guarding her newborn fawns, hidden somewhere in the woods between the houses. She fed alone in the yard every day, but, as Ade and I respected her nursery, we did not see her infant twins. In July, when the human commotion grew worrisome to her, she took the fawns to a quieter place; she returned with them in October, after the tourists had gone. On June 14, 1961, she had her twins in presumably the same place, took them away in July, and again brought them back in October, at which time three of her four offspring from the preceding two years, two yearling bucks and a doe, also came back.

There had been no violence here during Mama's sojourn with us, and the arrival of hunters in November, 1961, caught the little family group by surprise. As they stood in the yard on the opening day of the season, they reminded me of the little children of our dangerous times—beautiful, innocent, and totally unaware that death can strike from a distance.

A week later the bucks no longer grunted outside my window for corn and the young doe did not stare curiously in the windows. Mama, older and more experienced, fled with her fawns at the start of the shooting, and so these three survived. When they returned in December, they fed at night. Later they came by day but approached only when we were alone. Their return to some degree offset the loss of the others, but it did not take away the feeling that, by teaching the deer to trust us, we had taught them to trust all men.

Deer were seen around here during the summer of 1962 but Mama was discreetly absent. As the open season approached, Ade and I watched uneasily, for we are fond of Mama—but she did not return. Three days after the traffic and the shooting had stopped, I stepped outside at night to meet Mama's calm stare and catch the flashes of white as her new fawns bounded away into the brush.

Apparently Mama has learned that this place is no longer safe when other people move in the area, but in some way she senses that Ade and I are "safe" humans. We are touched by this and, in return, offer food and a quiet wintering place. If the lesson stays with her, when the snow is down next year and transient men are gone from the woods, they may return—two yearlings and Mama, with new twins to educate in good deer manners in our yard.

The Land Hunters

SUNLIGHT rayed through breaking clouds and the last flakes of the storm were falling as, knee-deep in fresh snow, I climbed from a blue-shadowed hollow up a glittering slope on our little side road. I paused on a hilltop, deeply conscious of the many miles of wilderness that stretched around me. In the absolute silence, I felt the beginnings of loneliness.

Abruptly, so near that it might have come from my side, the howl of a timber wolf rang out. The long, musical wail swelled, rising high and clear to warn the creatures of the sky, lowering and deepening to hurry small things into their burrows, spreading across the land to send the deer away to safer grazing, singing in the valleys and echoing from the hills, filling the quiet spaces of the forest. Ancient instinct tightened my muscles and raised the hairs on my neck. As I relaxed, my sense of loneliness went away. A wolf song has been company for man's solitary journeying since the days when men first trained wolves as hunting companions. I turned homeward, thinking of the clean, green land of yesterday, untouched by man-made good and evil.

Halfway down the narrow path from the road to the cabin I met a wolf, a big fellow all of five feet long, with

a might of power in his heavy muscles. His thick coat was shining gray, shading to lighter tints on his underparts and legs, with a beautiful wash of black flowing down his back and onto his bushy tail. His puffy cheeks gave the impression of a stifled grin as he looked at me steadily from interested, springwater-clear green eyes.

Apparently classifying me as harmless, he spread his forefeet, stretched his head up, and yawned. I saw his strong teeth and heard the snap as his powerful jaws came back together. Then this animal, whose kick could have knocked me senseless, whose jaws could have broken an arm or leg effortlessly, turned into the maple brush beside the path. While I walked past, he watched cautiously, then trotted away.

As I warmed my fingers by the stove, I recalled an account of a similar meeting in the early-settlement days. A gentleman met a ferocious wolf, which boldly blocked his way. The gentleman defended himself staunchly against this motionless menace, shaking his fist, brandishing his staff, shouting and grimacing, and finally calling various names, so that the wolf might think him one of a large company! At last the savage beast gave ground and slunk into some bushes, past which the frenzied gentleman escaped.

The interesting point is not that the gentleman and I reacted so differently to a face-to-face encounter with a timber wolf, but that the wolf, in both instances, behaved like a reserved dog on meeting a stranger. There was nothing about him to inspire the fear and hate that has led to his almost complete extermination in the United States. When the country was settled, timber wolves ranged from coast to coast except in the desertlands of California, Arizona, and Nevada. Now they are native only to the northern parts of Michigan, Wisconsin, and Minnesota, although a few may exist in the wilds of northwestern Montana and northern Idaho, and there are occasional transients from Mexico and western Canada. In 1958, it was estimated that the Superior National Forest harbored five hundred to seven hundred wolves; in 1962 the estimate was three hundred and fifty.

Any encounter with a wolf is a rare experience. Ade has seen them, watching from snowy ridges south of the road. We have both seen them running in their small family-packs along the frozen lake. But, in the warm months when visitors come here, the wolves are safe in faraway hills, devotedly rearing their puppies, which are cute—oversize feet, spiky tails, and all. A stranger could see a wolf without realizing that he had, because it takes practice to distinguish between a timber wolf, the smaller "brush wolf" that is a coyote, and certain dogs. The smaller and thinner coyote, found here in unimportant numbers, runs with its tail down, while the running timber wolf's tail streams behind it like a banner. The wolf's nose-pad is more than an inch wide, its legs are long, and its feet are very large, especially the forepaws, but I know of one instance where bounty was paid on a shot carcass before it was properly identified as the remains of an amiable old sled dog.

Until I moved to wolf country I had no idea how strong is the prejudice against them. Like a breath of Old Country werewolf superstition, or a racial memory of the caveman's competition against wolves for meat, it overrides modern knowledge and common sense.

All northern wolves, whether called gray, timber, arctic, lobo, or otherwise, are of the same species, and their pups may vary from black through red and brown and gray to any mixture in one litter—yet there is a local belief that black wolves are "Siberians," very dangerous and savage. This is as logical as the idea that red hair gives its owner a fiery temper. According to the very complete study *Wolves of North America,* there is no authenticated case of a man's having been killed by wolves in two hundred years of United States and Canadian history—yet new tales of hairbreadth escapes come out of the northland regularly. It is notable that those who encounter wolves always return to spin their yarns. Deer herds flourish and fail with the waxing and waning of their food supply over long cycles—yet hunters and bounty trappers insist that wolves cause deer populations to dwindle.

Wolves sometimes learn how easy it is to capture domestic livestock and pets, and it is obvious that they had to be eliminated from heavily settled areas; but, in wild places, there is no good reason for the bounties on their heads and very good reason for their preservation. When herbivores increase beyond the ability of the land to sustain them, they can cause serious damage to vegetation, before dying of starvation and disease. The wolf, by feeding his hungry family from these mammals, can be a valuable help in the management of his wilderness.

Our other wild dog is the red fox. It often strolls along a forest path or road lifting its black-booted feet daintily, waving its magnificent white-tipped tail, looking about with a disdain peculiar to foxes.

The cross fox is not crossbred between the gray and the red fox, but is a red fox with a tawny coat, marked with a black cross down the back and across the shoulders. The most beautiful fox I ever saw was coal-black, with a tailtip like a silver chrysanthemum. Both these color phases may occur in a single litter of four to nine puppies, as may also the name-giving red and the white-frosted black, known as silver. Less common are samson foxes, whose coats are woolly and lack the long, protective outer guard hairs, and bastard foxes, with normal fur but a color between red and black.

Summer people near us made the acquaintance of a cross fox that came regularly to them for his evening meal but did not neglect his daytime hunting along our shore. We saw him often during the following winter, sniffing at every hole and track in the snow. He disappeared in the spring, prob-

ably the victim of someone who coveted his fur or begrudged
him a duck or two.

His going did not make any observable difference in the
game-bird population, but there was a sharp increase in
voles, mice, and chipmunks, and a woodchuck took up
residence in our yard. Wild grasses and ground plants were
badly nibbled, produced little seed, and deteriorated rapidly.
An influx of owls and hawks restored the balance.

In 1955, Ade and I saw, in a rivulet of silt left by the
run-off after a heavy rain, a clearly defined, four-toed, claw-
less track—a cat track more than four inches long. The
cougar, long believed extinct in Minnesota, is the only
northern mammal that could have left such a mark. Soon
we learned that tracks had been seen in the southern part
of the state and that a mother and two bouncing, spotted
kittens had been observed near the place where we saw the
track. Then a visitor told us that "a big yellow animal with
a long tail" had crossed in front of her car on a road in the
southern part of the Superior National Forest. A few of
these magnificent mammals, withdrawn to hidden sanc-
tuaries, had survived against overwhelming odds.

The ranges of the southerly bobcat and the northerly
Canada lynx overlap in northeastern Minnesota; therefore,
the return of the cougar lets this wilderness boast of three
resident wildcats. They are valuable scavengers and controllers of herbivores. The lynx is particularly fond of eating
snowshoe hares, which, in overlarge numbers, can be a
serious menace to young trees.

Both the bobcat and the lynx measure about three feet,
with long legs, comically stubby tails, and admirable black-
and-white striped ruffs. The bobcat's coat is more heavily
spotted than that of the lynx and is often darker; the lynx's
softly shaded fur blends into the shadows of his boreal
forest. The lynx has long, prominent eartufts and the tip of

its tail is black all around, as though it had been dipped in ink; the bobcat also has eartufts, but they are short, and its tailtip is black on top only.

In 1957, I glimpsed a short-tailed cat looking down at me with eyes like golden balls from the branch of a snowy pine. As I tried to see details, it stepped with perfect poise to a higher branch and seemed to vanish in midair, like the Cheshire cat except that it left no grin behind. Only when I found its tracks, with their furry "snowshoe" rims, was I sure that our visitor was a lynx and not a bobcat, which lacks the snowshoe adaptation for traveling over deep snow. The smell of bacon rind tacked to a feeding bench had lured the recluse to our doorstep.

Such a sight is not so unusual now because 1962 has brought a large influx of lynxes. In a few months, more than a hundred were turned in to the game wardens for the $15 bounty. Many of those killed and bountied were emaciated. These usually retiring mammals have approached settlements, boats, and hikers, and have visited our feeding yard on two nights. Perhaps some disaster in the northerly part of their range has seriously reduced the numbers of the small mammals that are the lynx's main food.

There is a great size gap between the lynx that sometimes weighs above forty pounds and the pygmy shrew, smallest of all the mammalian world, whose weight is only as much as a dime. Three inches is a good length for these pointed-nosed, pinpoint-eyed bundles of energy, and two of the newborn young can rest on the eraser at the end of a pencil. The gray-brown arctic and common shrews look much like the pygmy, and are only slightly larger. These small creatures are under cover so much of their lives that they are rarely seen, and then only in glimpses as they dart across an open space. The short-tailed shrew, four to five inches long including an inch of tail, often frequents woodpiles containing old logs that harbor living grubs. These dumpy shrews, moving along as though on well-oiled wheels, remind me of animated fur coat buttons, dragging an inch of thread.

Shrews are insectivores, but will eat any meat, fresh or tainted, that they can get. Spoiled meat that is alive with larvae is a treat to them because they eat the larvae with gusto. They hunt incessantly, and may ferociously attack and kill animals larger than themselves. Consequently, they are reputed to be savage and gluttonous. However, their rapacious hunting is inspired by necessity, because they need to eat their weight in meat daily; their metabolism is so high that they may starve in a day, and often die of old age when not yet two years old.

A pygmy once trapped himself in a spaghetti-sauce can awaiting disposal in our kitchen. I tipped him into a pan, where, chittering angrily, he stretched all of two inches tall and, lifting his five-digited pink "hands," prepared to defend himself. I rewarded his bravery with a tablespoonful of the sauce that had led him into trouble. He ate it all, its disappearance within his tiny body verging on the miraculous.

Common shrews sometimes check over the logs in our wood-box. I dig grubs from under the bark and sit on the floor to handfeed our minuscule visitors. They dash at my hands, sniffing and bumping their noses against my fingers until each one locates and snatches a grub. I do not believe they are aware of me as anything more than background in their Brobdingnagian world.

We welcome shrews inside. They show little fear, even chattering and whistling when in the middle of the floor as though to warn us to keep out of their way! They do no harm. Mice do not appear in overly large numbers when the shrews are in residence, and invading ants vanish into the ever-empty stomachs of our littlest hunters. Small they are, but large is their service.

The family of the Mustelidae, or musk carriers, is represented here by the marten, fisher, mink, three species of
weasel, river otter, striped skunk, and possibly the wolverine.
All are furbearers and, except the otter and skunk, are persistent and efficient hunters of warm-blooded land animals.
The mink, however, spends so much time in the water that
I have included it, along with the otter, as a "water's-edge"
mammal. The slow-moving, stolid skunk, although it eats
some small mammals and ground birds, is largely insectivorous and vegetarian, and I have included it with the hibernating mammals.

The stone marten is a golden-brown, arboreal weasel, with
prominent rounded ears, a bushy tail, and a buffy or yellowish breast. It looks much like a small, long-bodied fox. The
fisher is a larger, darker edition of the marten, with relatively smaller ears and a more sinuous body. The wolverine
is a heavier, dark-brown relative that resembles a cross between a fisher and a small bear, with tawny stripings across
its forehead, along its sides, and over its hips.

These three mammals were exterminated in Minnesota by
fur trappers and hostile people years ago. The fisher has
made a strong comeback during the last decade and, in 1960,
indications of some thirty martens were seen in a remote
area where several were accidentally trapped. In 1961, a
friend of ours saw a wolverine that seemed to be weak from
hunger, crossing a road. The wolverine has been feared for
many years as though it were the devil incarnate, but Peter
Krott, the Austrian naturalist, raised some of the animals and
found them, when treated with kindness, so friendly and
gentle that they were excellent playmates for his small
children. Martens and wolverines have not yet visited us, but
would be most welcome.

Our three species of weasel are: the long-tailed weasel,
which may be two feet long and is found in most of the
states; the short-tailed weasel or ermine, whose winter coat

is the fabulous fur of kings; and the least weasel, only seven inches long and rare, which is the smallest living representative of the order Carnivora. (Smaller mammals that eat some meat are shrews and moles, both of the order Insectivora, and mice, of the order Rodentia.) All these weasels are milk-chocolate brown with white underparts in summer, and turn white with the onset of cold weather, although the long-tail is brown all year in southern habitats. The least weasel may be distinguished from the other two because it lacks their characteristic black tailtip. The long-tail has brown feet in summer pelage and the ermine has white but, in winter white, they cannot be distinguished by a casual glance, because the fur is similar and the size ranges overlap.

The weasel is the most beautiful and efficient mousetrap on earth, and does not deserve its ugly reputation. Its killing of more than it can eat is an attempt to procure food which may be stored and eaten later. Adult weasels need one-third of their weight in meat daily and the young, which number up to a dozen in a litter, one-half of theirs. That weasels are intelligent and resourceful enough to deserve anyone's respect was demonstrated to us by Walter, our adaptable weasel.

On a mid-December afternoon the screaming of blue jays told me that something was amiss in the yard. The center of the disturbance was the ermine we named Walter. Twisted into a fur rope, he was making acrobatic attempts to get suet from a cage hung on a cedar bole. Not having much success, he dropped down and bounded gracefully to the doorstep.

He was only a foot long and very dashing in his snow-white coat, accented by his black tailtip and the candy-pink lining of his rounded ears. His glittering black eyes peered past me. Then he sniffed, audibly and ecstatically, and licked the corners of his mouth with a fuschia-colored tongue. Walter was trying to tell me that, if I would kindly withdraw, he would be happy to go in and take care of the meat thawing in the kitchen.

If this was the crafty, ferocious weasel of ill-repute, he concealed his true nature adequately. He looked alert, eager, straightforward. He gave me an impression of intelligent interest that I had not seen before in any animal.

I wanted to know him better. Like all other wild things, his whole existence was directed by hunger and fear. By satisfying his hunger, I might overcome his fear. I tossed a bit of meat. He snatched it and flashed away, feather-light on his dainty paws.

Gradually we gained mutual confidence until he took meat from my bare hand with care and daintiness. When he occasionally mistook a finger for food, he pulled hard, then sat back and looked puzzled, as though he could not understand why this particular bit of meat was attached to me. Never did he show any tendency toward biting. From this point, I made no effort to train him, that I might learn how he would use his new and strange situation.

Walter was soon busy moving from the deep woods, a matter of finding storehouses and shelters under stumps, outbuildings, boulders, and brush piles near the house. He laboriously transferred one item of personal property, a meaty bone that was twice his length and weighed several times his three ounces.

He took suet and ground beef to each of seven storehouses. He liked frankfurters, potted meat, butter, and bacon, and a can containing remnants of boned chicken was a treasure to be carried home in his mouth. He arched his neck as proudly as a carousel horse as he held his head high to keep his pattering front feet clear of the dangling can. But he preferred red meat, especially moose meat, given us

by our Indian friends. Ade insists that Walter is the only weasel ever to eat anything as large as a moose.

My appearance outside after dark was a signal for Walter's head to emerge jack-in-the-box-like from one of his doorways in the snow, whence he hurried to the step to wait for a handout. When I did not go out, he attracted my attention by running up and down on the screen door.

Walter's life, from one of ease, became filled with deadly peril when his larger cousin and arch-enemy, the fisher, arrived. This fellow had all the suppleness and agility of the ermine, was forty inches long, weighed ten pounds, and could leap twenty feet—altogether a formidable adversary for little Walter. Even a female, perhaps two feet long and half as heavy as the male, would have been menacing to him.

The fisher began his campaign by systematically digging out Walter's storehouses and eating the contents. Then he concentrated on digging out Walter, who abandoned his open paths and found new ways to cross the yard: twisting routes under woodpiles, brush, any cover.

The fisher hunted at night, so Walter came daily for one large meal. When the fisher's night prowling did not yield Walter into his jaws he began to forage by day. Walter promptly began to come at night, scratching briefly on the screen and waiting for me under the woodshed. At last the fisher began to prowl at any time during the twenty-four hours.

Walter did not appear for four days and a fresh snowfall was unmarked by his small tracks. We decided sadly that our ermine had come to the violent end of almost every wild thing.

In the small hours of the next morning I was awakened by a touch on my face. Walter, trembling pitifully, crouched on my eiderdown, his face gashed from brow to nosetip and his right eye black and swollen shut. One of the fisher's raids had been a near thing.

After he had eaten ravenously, Walter went straight to the inside of the kitchen door—remarkable because he had previously approached it only from the outside. When I opened it, he crouched on the sill, tense and fearful. I stepped into the blue-bright moonlight and opened my robe to throw a shadow across the step. In this sheltering dark, Walter shot into the night. He came regularly while his wounds were healing, sliding through some small opening under the foundation (which we never exactly located), and timed his later visits with the hours when the moon-shadows darkened the doorway, but he would not stay inside.

Walter loved his freedom and his own wild way of life more than any pampered security we could offer him. Because of his courage and rapid adjustments, he still roams the forest, the only sign of his brush with death a twisted line of fur beside his nose.

The fisher that was the villain of Walter's story is, when considered from his own standpoint, quite as fine as the ermine. Everywhere wild creatures demonstrate that nothing is good or bad of itself, that circumstances and viewpoint too easily lead to attitudes of judgment that should have no part in the evaluation of the things of the earth.

The fisher retires to remote and dense evergreen wilderness, where men seldom stay long. Consequently, the little that is known of it in the wild state has largely been deduced from observations of tracks on the snow. Because the country in which our cabin stands has rocky hills for den sites, and an abundance of small animals for food, and because we try hard not to disturb the wilderness quiet, fishers drop in on us every now and then.

A fisher is an outstandingly beautiful animal. More than

a third of its length is fluffy, tapering tail. Usually its coat is so dark a brown that it seems black, with long, upstanding, pale guard hairs about the head and shoulders, that give it a frosted look. Sometimes the fur is a light or reddish brown and, in 1960, a cream-colored mutant was accidentally trapped a few miles from here. The fisher's body has the elongated form and agile grace of the weasel and mink. Its legs are short, especially the front ones, and powerful haunches foretell great leaping power. Its face is the slender face of the weasel, with the same interested eyes, small ears deep in fur, and sensitive pointed snout. Usually there is a small white throat patch.

By night the fisher is as fearfully exquisite as a creature out of dreams. Moving about in the cold light of the stars, moon, or aurora borealis, it is a mysterious, fluid part of the half-dark. The frosty hairs that give it daytime fluffiness are invisible and, smooth and sleek and sinuous, it flows and poses, a shadow darker than all other shadows, its eyes like emeralds exploding into flame. It glides in the unearthly beauty that belongs to the untamed land and its children.

My first daytime view of a fisher came when I was sitting on the step, admiring the beautiful stripings and rich red wash on the rump of an eastern chipmunk that was standing in my left hand and gobbling corn out of my right. I sensed movement in front of me. Crouched not ten feet away, a big male fisher glared at the little animal in my hands.

Very steadily I stood up, carrying the chippy along. Startled, it almost jumped away, but caught sight of the fisher and collapsed in fear in my hands.

When I moved forward a step, the fisher rose on his haunches and hissed, but he gave ground and leaped to an observation position some five feet up the trunk of a small tree. He clung there much the way a black bear does—one foreleg wrapped higher than the other while he peered around the bole to see whether he might safely come down or should seek shelter higher up. When I made no other move, he inched himself down backwards—although they can literally flow head downward—and trotted away without haste, giving me an occasional over-the-shoulder glance.

The quivering of the frightened chipmunk stopped. I looked down to see it calmly stuffing its cheek pouches as though nothing had occurred.

During the past eight years, Ade and I have seen many fishers and even persuaded some of the more hunger-driven to feed from our hands. Always they took food with care not to touch our fingers. However, a trapped or cornered fisher will defend itself and its freedom ferociously.

I will accept the fisher as an animal that bears within it some of the same unquenchable spirit of wildness as the timber wolf, an animal that adds joy to my days because of its beauty and grace, that sometimes stirs in me an eerie whisper of ancient hauntings as it moves like a bodiless shadow in the moonlight or pads ever so softly across the roof in deep night.

Mammals of the Water's Edge

FALL is the best time to explore the edges of the brook west of the summer house. Then the water that gurgles between the tussocks has drained underground, and the mosquitoes are almost gone. The brook has dwindled, but not dried up, and the newly deposited patches of fine silt show tracks as sharp as fingerprints on glass.

I knelt on a twisted cedar root to examine some interesting markings. Mouse tracks? No—shrew. The prints were very like those of tiny human hands with the palm elongated between the thumb and the wrist; mouse paw marks are not so handlike, and there are only four toes on the front feet. The gray water shrew that had made the tracks scuttled from under the root below me. He stepped from the bank and ran underneath the water along the brook bed. A current dislodged him and he rose like an inflated toy but, apparently not in the mood for water sports, he overcame this extraordinary buoyancy by vigorous swimming and returned to the sandy bottom.

A large worm, of the shade of pink that looks like something freshly skinned, crawled over the rim of a shadowed hollow in the brook bed. The shrew pounced. The worm twitched and jerked, but the shrew—about six inches from pointed nose to tailtip, less than half as long as the stretched-out worm—hung on and made his way sideways and backwards to the far bank. So violent was the struggle that the air entrapped in the shrew's fine, dense fur—the secret of his buoyancy in water—was dislodged, and streamed upward like gas bubbles in soda pop. Disturbed silt clouded the roiled water. Then, with furious splashing at the far bank, the shrew dragged his still-resisting dinner under a clump of ferns.

I found a dozen of the worms, coiling and slithering in the hollow, and lifted one out of the water. It was a segmented relative of the earthworm, with a thick band near its posterior end. Suddenly the chilly creature stiffened, shortened, and broadened to an inch wide, arching so that its curved rear with the swollen band gave a fleeting impression of an eight-inch cobra with hood expanded. Startled, I let it go back to its pool-crawling.

The shrew was very light in weight, but its tapering snout had a mouth fitted with sharp, hooked teeth. No doubt its luckless prey, in spite of superior weight, greater size, and considerable strength, was being rapidly devoured beneath the greenery.

I went on downstream toward the place where Hoppy, the three-footed mink, had his den in the bank. There is much that is breath-taking about a small rocky stream, when you take time to look closely. Caves the size of thimbles, inch-high rapids, even a thread of a waterfall that caught a newborn rainbow in its silken fan of mist. By the time I reached the stone where I might meet Hoppy, the frozen liver I brought for him had thawed. I was washing my bloody fingers in the brook when he poked his brown face from between two rocks and sniffed the delectable odor. While he ate, and cleaned his abundant whiskers, I waited to see if this was one of his playful days.

It was. Gradually he approached one of my hands, hissing as he stalked, arching and curving his tensed and sinuous body. This was the beginning of a game that one might call "let's bite fingers." The object on my part was to wait until he sprang and try to jerk my hand out of reach. Needless to say, I was almost always defeated. This did not result in so much damage to me as one might think. Now and then Hoppy snapped too hard, but he usually caught a finger lightly enough not to break the skin.

I am proud of my friendship with Hoppy, because he had chewed off his right forefoot to escape from a trap and did not look on the human species as acceptable in any way. The first time I saw him he was reared up on his hindquarters, staring as though hypnotized at Creampuff, a white hen fastened outside the chicken yard because she had a seriously pecked comb. Creampuff was gawking as fascinatedly as Hoppy. It was a picture—the fluffy white hen with rose-colored comb and bright-yellow beak, and the sleek, opalescent-brown mink against the heterogeneous green. I expected disaster. Instead, Hoppy turned away slowly and limped toward the stream, with many puzzled backward glances.

For the next two years I put out meat near the brook and waited with that frozen patience required in making the acquaintance of any suspicious wild thing. A big snowstorm finally made finding food so difficult that Hoppy, spitting and growling and snarling, crept up and took a piece of meat from my boot toe. From here on, our association advanced rapidly, and he behaved much as my weasel acquaintances have—climbing onto my lap to take food from my fingers.

He apparently was a confirmed bachelor; he did not roam during the April mating season, and no lady mink appeared with whom he might share the responsibilities of rearing perhaps half a dozen young. I doubt that he would have

been so trusting as to let me feed him almost every day dur- MAMMALS
ing the summer if he had been the father of a family. As it OF THE
was, he let me share the edge of his hidden world for four WATER'S EDGE
years, but he never approached the house. 159

Feeding him thus required a walk through the woods in all weathers. I can thank Hoppy for the feel of frost needles whipped against my face by the wind, the sight of rain-splash patterns in dark pools, the shiver of danger that comes when a blinding storm obscures trees at arm's length—and for my aversion to mink coats.

A visitor, boat-riding with Ade and me, stared past my shoulder and turned pale. "What *is* it?" she quavered. "It looks like a big snake!" Not expecting to see a relative of the Loch Ness monster in our waters, I gaped at a series of dark loops moving toward the shore. Once there, it emerged as six river otters, one following directly behind the other. They galloped in a string to the land end of a log, pranced to the other end that overhung the water, and dived in, one after the other, as gracefully as Olympic champions; whereupon the "sea serpent" reassembled and vanished around a rocky point. (I have always wondered whether they were playing follow-the-leader as a change from their bank-sliding and underwater pebble-tossing games.)

This water-going cousin of the weasel is several pounds of exuberance dressed up in a sleek, brown coat. It looks much like a big mink, with a broader snout and webbed feet, but spends more time in the water. It can be confused with the lighter-weight fisher, which also swims on occa-sion. Ben Ferrier, a writer and lecturer who has traveled over much of the wild north, told me that he once watched two "otters" swimming toward him. As they emerged on shore he was surprised to see the thick, flowing tail of the fisher instead of the smooth, heavy-based one of the otter.

Otters, except females with young, are wonderfully friendly and make endearing pets, sometimes even talking to one in otter language. We have refused several offers of young otters for the same reason that we kept to ourselves Hoppy's presence by the brook. Too many tame fur-bearers die at the hands of covetous men, as Gavin Maxwell's Mijbil died, at Camusfeàrna by the bright waters. Mr. Maxwell's story of Mijbil and Edal is the story of all good otters everywhere.

In this part of Minnesota, as in other places, the otter's handsome pelt was almost its undoing. Heavy trapping reduced its numbers seriously but, under control, the jubilant mammals are making a comeback. Mother otter usually bears two or three pups—occasionally only one or as many as five.

Emil Liers has reared otters at his home on the Mississippi River, near Homer, Minnesota, since he acquired two young ones in 1928. He travels throughout the United States, accompanied by his pets, lecturing on their ways to schools and scientific and sportsmen's groups. He has debunked the idea that otters are fish destroyers. He found that otters fed exclusively on fish become seriously ill. Only a few fish are taken by them in the wild—their principle foods are frogs and crayfish. They also eat snakes, snails, insects, worms, and suchlike, garnished with some pond vegetable matter. Since crayfish feed heavily on young fish, the otter is a benefactor of the fish population.

The raccoon makes as nice a pet as the otter. Its black bandit mask with white above and below, its gray-brown body, and its ringed tail, alternately black and yellowish-white, are very familiar. Ade has a special fondness for these nice mammals.

When he was in his early teens, he stopped on a Chicago street to stare with fascination at a tall, swarthy man who was leading on leashes a raccoon and a porcupine, each about two feet long, barring the matter of tails. Both sat up and offered polite paws when their master introduced the raccoon as Achmet and the porky as Abou. As if this were not wonder enough, it turned out that the man was Abdullah ben Ephraim, a circus sword-swallower. He explained that Achmet and Abou "guarded" his swords during his performance. The three performers were staying at one of those old-time theatrical rooming houses, now sacred to the memory of ventriloquists and song-and-dance men. The acquaintance flourished; Abdullah lent Achmet to Ade for three weeks, feeling that a raccoon was safer to put out among strangers than a porcupine.

Achmet was the perfect house guest. He selected the secluded corner where his sandbox for sanitary purposes should be located, and chose a medium-size round dishpan as a "pond" for washing his food. Raccoons may eat unwashed food when water is unavailable, but Achmet had firm convictions about this. He had problems with oatmeal, which he liked very much but which, unless much dried out, fell to pieces when he dunked it. He spent his spare time sitting on the kitchen table, happily guarding the bread knife.

Raccoons are said to live near lakes and streams in all of the United States except certain western regions that lack water, and possibly northern Maine, but our rugged forest is not raccoon country. Stories of raccoons here are usually nth-handed hearsay, but in January, 1961, the game wardens accidentally caught a raccoon in a livetrap set for fisher. Raccoons retire to sleep during very cold weather, but their temperature, pulse, and respiration remain normal. Perhaps the captured animal had wandered up from more southerly woods and, not being well enough supplied with food to last through the long winter, came out in search of more.

Muskrats are rodents that resemble foot-long, overfat mice, with dark fur and long guard hairs, and tails that might be gray snakes, squeezed flat by pressure on both sides. I think their old Algonquin name, musquash, becomes them.

Although muskrats are present in northeastern Minnesota, I have seen none near here. I remember them and their "grass huts" along the slow-moving streams in Ohio. These were ideal locations for them, where cattails and rushes in shallow water could be fastened together to form conical huts that would be gradually enlarged and rounded until, old and much repaired, they resembled beaver lodges. Nearby were meadows that supplied food. Muskrats eat hugely of anything vegetable, and have a special fondness for corn in the ear.

The marshes were dotted with not only muskrats' lodges, but also their grassy feeding shelters and rafts. The first mammals I saw born were five bare and ratlike muskrat babies that squeaked into life on one of the rafts near the shore. Mother, gripping the belly skin in her teeth, carried them one by one to her lodge on a midmarsh island, where snapping turtles that lived on the muddy bottoms would be less likely to find and eat them.

One winter I accompanied a neighbor-boy named Lewis to the muskrat marsh. We watched one of the animals swimming beneath clear ice. Suddenly it breathed out a big bubble of air that, after an interval, it breathed in again. I think that the water had removed carbon dioxide and restored oxygen. When we saw a second muskrat with its bubble, Lewis thumped on the ice and the bubble was broken up. He saw that the little animal could not breathe in the resulting small bubbles and was drowning, so he broke the thinnish ice with a pole, incidentally dropping both of us kneedeep in the cold water. The muskrat popped its head out for air and the shivering young naturalists ran all the way home.

Beavers, equipped with nose and ear valves to seal out water and with large lungs and liver to oxygenate their blood, can stay underwater for fifteen minutes, but they also make emergency use of the bubble-under-the-ice techniques. They are larger than one thinks, often weighing fifty pounds, with fat old males sometimes reaching a hundred. The humped brown body, with its valuable waterproof fur coat and gray, scaly tail in the shape of a paddle, cannot be mistaken for any other mammal. The beaver's construction work is remarkable and unique, and in a daytime encounter one is more likely to see the workings than the wary builder.

Last summer Ade and I came on a beaver dam across a small stream that flows into a nearby lake. There was a fine pond behind it, and a large lodge near the pond's edge.

The purpose of the dam is to assure a depth of water sufficient to allow underwater entry to the lodge after the pond is frozen over. The lodge is built on a flat mud bank or on an island, natural or constructed of mud by the beavers. Tunnels are dug through this foundation from the water to the living room. The lodge's walls are of rocks, logs, debarked food sticks, and mud, built upward and inward to form a circular hut. The outer five or six inches of the perhaps three-foot-thick walls are of mud plastering, which freezes rock solid in winter, except at the top. There, thin plastering or none at all allows the inhabitants' body heat to melt the snow and permit ventilation. The living chamber is about three feet across and two feet high, with its lowest level, where entering beavers shake water from their fur, an inch or two above the water level in the entrance tunnels. One of these leads to the underwater food supply, and others may lead to the dam, or give extra escape routes if the lodge roof is attacked by a bear or other large predator. The living quarters are raised and dry, sometimes carpeted with chips, although the adults often sleep on the bare earth. The one to

six open-eyed and softly furred babies, usually born in May,
rest in a nest of grass, leaves, roots, and twigs, which can be
eaten when the young are stronger.

Beavers consume the bark of a wide variety of deciduous
and coniferous trees and shrubs, but have individual prefer-
ences. In warm weather they add terrestrial and aquatic
plants as greens. Near the lodge we had discovered, Ade
and I saw a goodly number of cut aspen and birch logs of
small size, weighted beneath the water with stones for winter food. This puzzled us, because only jack pines grew near
the pond's edge.

Then we saw the canal, about eighteen inches deep,
reaching some fifty feet across the flat, sandy ground. Earth
that was removed when the canal was dug banked its edges
slightly. The flat area may have once been part of the bed
of a large natural pond, since drained by the stream, and
now partially refilled by the waters of the smaller beaver
pond. The canal was filled by flooding from the beaver pond,
although ground water also drained in. It ended at a steep
but shallow bank, possibly the edge of the earlier natural
pond. From this point a second canal, higher than the first
and closed by a low dam that acted as a sort of lock, extended to a grove of aspen and birch. We could not see the
source of the water in the upper section of the canal, but we
think it came from a spring just below ground, for these are
common here. In the grove, there were many stumps with
conical tops formed when beavers gnawed round and round
the trunks until the small remaining center sections snapped
and the trees fell. (These energetic lumberjacks cannot control the direction of a tree's fall. Sometimes they are crushed,
and occasionally the tree jumps off its stump and, remaining
upright, pins the unfortunate beaver's tail to the ground, the
pointed, gnawed base of the trunk driving into the earth
like a spike.) At any time except in serious drought, these
canals would be great labor-savers in bringing harvested logs

to repair the dam, or to be stored for food. In wet weather, the higher canal would spill its excess water through the lower one into the pond, and thence over the dam into the stream bed.

This dam was only about twenty feet long and two feet high, a small one, as are many in rough land. In flatter country the dams may be much larger, but are generally less than three hundred feet long, and not more than five feet high. One of the longest on record, on the Jefferson River near Three Forks, Montana, measured 2,140 feet. Really large dams may rise to twelve feet in the central part, and have bases fifteen to twenty feet thick. Generations of beavers worked to complete these imposing structures. Beavers were once found in plenty from coast to coast and from the Arctic to the Mexican border. Their workings flooded thousands of acres, and their big ponds dotted the continent before the white man came.

As the water slowed down behind the dams, silt dropped out. Gradually the ponds filled and became so shallow that the beavers moved on to build new dams and create new ponds. Slowly the abandoned dams weakened, while silt still collected behind them. One by one they began to leak and drained the water from the rich, flat beds of the ponds, where grasses sprouted and held the soil. After many years, forests grew in some of these places. Meanwhile, the newer beaver dams were collecting more of the rich soil that eventually attracted white settlers.

Then trappers broke trails into the wilderness to get the beaver pelts; homesteaders, roads, and civilization followed. The beaver was trapped almost to extinction.

This, as is common with upsets in the natural scheme, led to disaster. Large numbers of untended beaver dams broke out almost simultaneously and there were few new dams to control the released water. Men who wished to farm the still-flooded land blew up many of the remaining dams, much as men are draining marshes today. There was great loss of water from the land.

When rain fell on the newly bared silt, unprotected by natural growth, erosion followed, and ruinous floods. Added to this was the loss of water and soil from forest lands that were stripped of their trees, and from prairies that were denuded when the buffalo grass was plowed under. The result was calamity, the causes of which were not seriously considered until floods had repeatedly ravaged the valleys of the Mississippi River and its tributaries, and the dust storms of the 1930's had carried away much of the topsoil from the plains.

Reforestation, contour plowing, and other reclamation measures were put into effect. In 1938, the United States Government recognized that, although beavers may be a nuisance when they fell fruit trees or build dams that flood roads and fields, they are natural conservationists whose unappreciated ancestors contributed much of the nation's usable water and fertile soil. In Idaho, a thousand beavers were transplanted to areas where erosion control was needed. The experiment was a success.

Since then, many nuisance beavers have been relocated. These tireless construction workers have been responsible for raised water tables, lush meadows, and abundant fresh water in areas that were near-desert before their coming. Arizona, especially, has benefited from the removal of beavers, whose dams interfered with the all-important irrigation systems of the lowlands, to desolate high country. These beavers, enticed into wire cages with juicy aspen bark, were transported on horseback in gunny sacks kept wet to cool the animals and to prevent drying, chapping, and cracking of their hind-foot webs. When they were freed at the stream that was to be their new home, a pile of freshly cut brush was waiting to give them shelter and to encourage them to start building. This system satisfies men and thirsty crea-

[handwritten marginalia: beavers are GREAT conservationists]

tures, both wild and tame, while it brings greenery to the once-arid land and gives homes to the beavers, those remarkable engineers without portfolio whose value in flood control probably would have exceeded by far the private fortunes made from their fur in the past.

The Hibernating Mammals

IT SNOWED LAST NIGHT, and the branches are so loaded that white bosses reach from tree to tree. The ground is buried three feet deep. The wind that cleared the sky brought a temperature of thirty-five below zero. The only tracks this morning were those of a snowshoe hare and a deer, both of which came for corn. Now a few red squirrels are feeding but will soon take shelter along with a scattering of jays and woodpeckers.

Chickadees, puffed into feather balls, search out small crumbs from the snow or pick fragments from the suet feeders. When they seek shelter, the yard will be empty and silent—apparently lifeless. I think, with passing unbelief, of the thousands of living things hibernating all around me: minute worms, wrapped in silken blankets; pupae of many sizes and shapes; butterflies, asleep with their rainbow wings folded; frogs and toads and knotted skeins of garter snakes, lying as still as the stones and earth under which they have sought cover.

Many mammals are hibernating, too, but their sleep is not always so deep as that of insects and cold-blooded creatures. In true hibernation, the pulse rate slows, the breathing becomes infrequent, the metabolic rate and body temperature drop, and other bodily functions cease.

The little jumping mouse goes into such a dormant state that, if disturbed, it will lie in the hand as though dead. Why these mice hibernate and others do not is puzzling. However, the jumpers are not closely related to true mice; their nearest American relative may be the very different porcupine. This does not solve the puzzle, though, because porcupines are wide-awake all winter!

Our yard has two species of jumping mice, the woodland and the meadow, which are indistinguishable except at close quarters. Hopping through the grass, they may easily be mistaken for the brown form of the leopard frog, although the mouse's leaping is smoother and more continuous.

I once had a fine, sunlit view of an adult woodland jumping mouse that was hemmed in between two large stones. It had a brown back, bright-tan sides, yellowish back edges to its large ears, and pale-gray belly and feet. Its tail, about six inches long and almost twice the length of its head and body, bore a prominent white tip that is the clearest difference between this mouse and its meadow relative, whose tailtip has only a few white hairs.

One spring I found a very young jumper nearly dead in a puddle of cold water. When he quieted in the cup of my hand, I gave him a drop of warm milk, but he poked his nose into the liquid and almost strangled. I finally diluted canned milk with an equal part of warm water and spread a thin film of this on my palm. The baby mouse quickly licked up the accumulation in one of the creases of my hand and toppled sideways, fast asleep. At this moment there was a knock at the door and I spent the next half-hour holding a sleeping mouse while I talked to a neighbor, who plainly thought I had been in the woods too long. _ha!_

The infant mouse ate bits of wet graham cracker, but I could not teach him to take milk from any surface but my hand. This meant that I had to feed him every two hours, because his stomach was a marvel of smallness. I settled him

in a box on the kitchen floor with some shreds of cotton, and
he did very nicely, growing plump and furry while I turned
pale and haggard from broken sleep. But I forgot that he
might be reached by an enemy. I went to feed him one
night and saw a short-tailed shrew slipping out the crack
under the door. The aggressive shrew, three and a half
inches long without his inch of tail and weighing perhaps a
half ounce, had killed and eaten the frail young mouse,
whose head and body measured two and a half inches.

When I was small, I believed that certain mice had
wanted so much to fly that they had grown wings and become bats. I admired them greatly for this imaginary ac-
complishment, and brought confusion into the lives of
harvest mice nesting in a vacant lot by searching for a young
mouse that might be in the process of wing-development.
Bats, however, are unique in the animal kingdom, the only
mammals with the power of true flight.

Just why these creatures, which are enormously valuable
in insect control, should be thought omens of menace is a
question. That they would commit suicide by diving into a
snare of human hair is absurd and, although they have parasites, these are most often found on tropical species and are
peculiar to the bats themselves. The mouse-sized vampire
bat that carries rabies and certain diseases of livestock, is
confined to the tropical Americas. Rabies is rare among the
species of the United States and Canada, which eat only
insects and seldom come in contact with warm-blooded animals from which they might contract the disease.

There is much that is mysterious and wonderful about
bats and much that is not generally known, in spite of their
large numbers and their occurrence on every major land
mass except the polar regions.

The hand has been formed into the wing, with membrane
extending from the elongated finger bones to the forearm,

side of the body, and hind leg. The thumb remains free and is hooked for hanging. (Bats hang upright by their thumbs and upside down by the five modified toes of their hind feet.) The wings are truly almost "skin and bone." Their membranes, and the interfemoral membrane that stretches between the hind legs, are composed of two layers of skin, with nerves, hair follicles, blood vessels, sweat glands, and small bundles of muscle fiber between them. The wings vary in shape: some, long and relatively narrow, are suited to distance flying; others are shorter and wider to give greater mobility.

In contrast, a bird's wing is an adaptation of the whole foreleg, attached to the body only at the "shoulder" joint. The end section of the wing is the much modified "hand," with the remains of the "thumb" and first two "fingers" visible in the skeleton. This bony structure is lightly muscled; the powerful muscles that move the wing extend to it from the breastbone. The wing's expanse is formed by large feathers that rise from the back of the "forearm" and from the "hand."

Most bats are insectivorous, taking their prey in flight, the interfemoral membrane serving not only as an aid in flying but sometimes as a scoop to capture insects. Bats outside of the United States and Canada show many diet variations with accompanying physical modifications. Some eat fruit and small animals, and have both prominent canine teeth for piercing and tearing and well-developed grinders for crushing and softening their food. Some subsist largely on flower nectar and pollen; their snouts and tongues are long, the latter sometimes equipped with a sort of brush, and the teeth are poorly developed. There is a Central American species that, while flying, gaffs fish by means of specialized hind feet.

Most northern bats migrate, as hibernation is possible for them only where they may live in caves, mines, or other sheltered places where temperatures are near or above freezing. Not far from our cabin numerous small bats (probably little brown myotis) emerged for several springs from crevices in a cliff. Winter temperatures had reached fifty degrees below zero, but heavy snows had covered the cliff face and kept the bats snug. In the winter of 1957–1958, the snows were unusually light; only a few bats could be seen the following spring.

The hibernation of bats, when it occurs, is a special situation. The bats rest with lowered metabolism not only during winter cold, but also during periods in warmer weather when unseasonable chill, storms, or food shortage makes living difficult. This contributes to their longevity. Banded bats have been recovered after twenty years—an astonishingly great age for so small a mammal. An understanding of these controls of vital functions may have more significance for men than a satisfaction of intellectual curiosity. We stand at the threshold of travels into outer space—journeys whose length may be limited by man's lifespan. If man can learn the secret of controlled dormancy, what horizons might some day open for him?

Bats differ from other small mammals in that births are often single and, in most species, occur only once a year. The gestation period is very long, considering the animal's size—from sixty days in the little brown myotis, up to eight or nine months in large bats—and the length may vary with metabolic changes, resting periods, and outside temperature so that development is slowed down in unfavorable conditions. The young are large. The little brown myotis often produces a youngster whose weight at birth would correspond to that of a fifty- or sixty-pound human baby!

The mother bat bears her young while hanging upward, supported by her thumbs. The newborn bat is caught in its mother's interfemoral membrane and immediately climbs up her body to reach the nipples, where it clings even when she flies after food. As the baby grows rapidly, the mother carries an astonishing burden. When the young are able to hang safely in their home loft or tree or other secure place—after only a few days in some species, and about two weeks in others—the mother hunts alone. At about three weeks, the young bat practices fluttering until it is able to fly and can hunt for itself.

Bat individuals catch hundreds of moths and other soft-bodied insects in an hour. The echo-location by which they navigate and locate their prey is one of their most remarkable specializations. When flying, they send out regular pulses that speed up when an insect is detected, and increase to a buzz when they near their prey. These supersonic vibrations are in the radio-frequency range above the limit of human hearing. The bat's sonar is effective in highly noisy areas and a bat does not confuse its own pulses with those of other bats. Echo-location is not just a crude device for hunting and preventing collisions. Experiments at Harvard University have shown it to be billions of times more efficient and sensitive than man's best radar and sonar, and so exact that bats can feed and avoid obstacles in complete darkness. They can distinguish objects by shape, and can tell whether bars used as obstacles in experiments are vertically or horizontally placed. They avoid close-spaced vertical bars because of their wingspread limitations, but, by timing wingbeats, can pass between horizontal bars of close spacing.

By echo-location and mobility in flight, bats avoid enemies, such as owls. Although they drag themselves awkwardly by their wings on the ground because their hind legs are so specialized for hanging that they may not bend forward, they can get into the air so rapidly that they are seldom captured by night hunters. Their deaths are often due to old age.

Bats begin to hunt at dusk, but the chances of identifying one on the wing are slight, especially against a forest background where the bat is a shadow fluttering against

varying darkness. I once recognized a red bat swooping on a moth. The insect was perhaps four feet away from me and the bat flew directly toward me in a beam from a lantern. Its reddish color was plain and its small, almost fat face, with ears rounded like shells, was as attractive by human standards as bat faces can be. I opened the door on another occasion as a large gray spider, dropping from the lintel on its silk, folded itself into a protective ball in front of my nose. As I leaned nearer to see the arrangement of its jet-bead eyes, a small brown bat came into the light of the doorway and took the spider without pause in its flight. It was so near that I heard the snick of its teeth. This bat was either a little brown myotis or a keen myotis. These two species can be positively separated only in the hand; the little brown's ears reach to the nostril when laid forward, while the keen's ears extend a sixteenth of an inch beyond the nose.

On summer days the relatively large hoary bats hang upside down in small groups on high branches. Their frosted backs, yellow to mahogany, and their buffy throat patches are efficient camouflage, and the bats look like bits of sun-dappled bark to the unaided eye. One of them lighted on my shoulder while I was watching an August night's aurora. I flashed a light on it long enough to identify its distinctive coloring. By the faint glow from the sky, I watched it fasten itself firmly to my shirt by its left foot and work at one of its ears with the toes of the right. Once its all-important means of navigation was satisfactorily clean, it fluttered away.

I doubt that this bat mistook me for an inanimate object, because I was walking and carrying a light when it settled on my shirt and it remained indifferent to the movements in my shoulder as I craned my neck to watch it. Nor was there any indication of an accidental collision, which might have resulted if its ear problem had seriously affected its echolocation. Perhaps, with its "sonar" needing attention, the bat had simply settled at the nearest emergency landing place.

If I am lucky enough to be in the right place at the right time, I may see the other two species found here: the small silver-haired bat, so brown it is almost black, with white-tipped hairs down the middle of its back, and the big brown bat, distinguishable by its size and black membranes.

These interesting creatures are so retiring and nocturnal that they offer few opportunities for close observation outside of captivity, but opportunities do come, and in the most unexpected places.

In the Chicago Loop several years ago, I found a little brown myotis collapsed on the sidewalk outside of the office building where I worked. He moved slightly when I picked him up but was so parched with thirst that his mouth was open and his tongue swollen. An interested soda clerk gave me some paper straws and, in my office, I managed to give the bat several drops of water from the end of a straw. He moved, half-opened one dull eye, tried to stretch a folded wing. I put my patient in a large desk drawer and, after several more drinks, he was able to crawl feebly.

I took him home in a ventilated box and hung him by his feet on the edge of a shelf in my dressing closet. When dusk came, he had recovered enough to flutter. I filled the basin in the adjoining bathroom and opened a window from the top. The bat flashed gracefully back and forth, scooping up water in flight with his tongue, before he found the opening and went into the night. I left the window open and in the morning my bat was again hanging from the shelf.

He remained with me throughout the summer, sometimes staying away all night but often hunting only during the hours after dusk and before dawn, with a rest in the closet between. During a plague of May flies, when the streets and air were literally filled with the ephemeral creatures, he was gone only an hour—sometimes less—at a time, and spent most of his night resting indoors. It seemed that the duration of his hunting trips was determined by the availability of suitable food.

My bat was very clean and often spent twenty minutes or more grooming himself. He licked his wings and fur systematically as far as he could reach with his long tongue. He cleaned his ears and face thoroughly with his moistened toes. He allowed me to handle him after the first few days, and I examined him closely. This dashed any hopes that he might be a female and present my household with a young bat, but assured me that he had no signs of vermin.

He tolerated being stroked and enjoyed flying around the living room, thus creating near-panic several times when I had guests. I have always hoped that, when he disappeared in the fall, he went to some warm and moth-rich place to spend the winter.

Another light winter sleeper, once common here but now rare, is the pretty black-and-white striped skunk. Automobile traffic has taken a heavy toll because the skunk is so well protected by its odorous spray that it does not understand the need to make way for anything. Rabies is said to have killed many, but we have seen no rabid animals of any species here.

Lodgekeepers and summer residents are inclined to kill any skunk that makes the mistake of approaching a human dwelling, but summer people near us took a more reasonable attitude when a family settled under their cabin. When the skunk *pater familias* stood foursquare in the path, our neighbors used roundabout trails. So quiet and odorless were the skunks during their four years of residence that no outsider discovered their presence. Even our neighbors' dog raised no problems, for she had assaulted a skunk when she was a puppy, with results she did not forget.

A mother skunk, searching for insects and berries, will

mildly lead her children in a small parade, or a lone male will stroll calmly past a human observer. But let the human approach the skunk or appear to challenge its right of way and amiability goes out the window. The skunk faces his enemy and pounds his front paws on the ground. If the enemy does not withdraw, the skunk whips into a U-shape, raises its gorgeous tail out of harm's way, and sprays, simultaneously or singly, twin streams of musk from glands located at the top of the hips.

The skunk's musk does not cause blindness, but sets up a strong temporary irritation of the eyes, which gives the short-legged sprayer time to escape. The odor causes no serious harm, either, but is unsurpassed as a producer of embarrassment and inconvenience. And how it lingers! Scrubbing with gasoline seems to help, if one has a skin tough enough to stand such treatment. As to clothes, the simplest procedure is to bury or burn such reminders of human folly.

One morning last June I noticed that the pea vines were disappearing from inside the garden fence, and Ade pointed out some pole beans stripped of their leaves to a height of four feet. The fence was a tightly stretched, three-foot barrier of chicken wire. Hares could jump it, but they cannot climb poles. Chipmunks could both pass through and climb, but it would take an army of them to eat such a quantity in so short a time. As we wrinkled our brows, a rounded hump of gray fur sat up just outside the fence and contemplated us with grave suspicion. A woodchuck—and he had come unerringly to the only garden within miles!

Ade named him Gregory for a groundhog character in one of my children's stories, then mended a break in the wire which looked likely to be the newcomer's entrance.

Just before sunset, I saw carrot tops waving where no breeze blew. As Ade and I approached, Gregory sat up in

the carrot patch, chewing his last-plucked mouthful, and studied us with the calculating squint of a bettor figuring track odds. Deciding they were not in his favor, he tried to dive through the fence.

He stuck at the hips. He clawed frantically as he tried to pull himself through, while his fat rump waved and his stubby black paddle of a tail almost twitched off. At last he attacked the wire with his teeth. I am convinced that he thought he had cut the wire because almost at once he gave a complicated wriggle and was through, lying panting on the ground while he kept a bright and wary eye on us. Slowly, as though he hoped we might not notice, he crawled an inch, another inch—then bolted into the ferny ditch at a lumbering gallop. The fat, two-foot-long animal had huffed and puffed until the wires of one of the two-inch hexagonal spaces had spread enough to let him through!

We considered Gregory's discerning taste and huge appetite before we turned in. The peas were through bearing. The carrots were stunted by the drought. But the beans were doing well—and our Kentucky Wonders grow fifteen feet tall and bear pods that weigh in at sixteen to the pound. A fence of window screening, with heavy stones on both sides of the bottom, would stop him—if he did not learn to dig under it. I listened to a scrabbling at a trellis on which I had been coaxing an exotic vine to grow. Oh well. With the drought and the cool nights it probably would not have bloomed anyway.

When Gregory discovered our piles of corn he spent considerable time relaxing outside the door, lying down while he ate brunch and dinner. His fur was thick, glossy, and well-groomed; he had black feet and tail, reddish sleeve-guards and vest, and leaden head and back, the latter heavily grizzled. His bright, black eyes and bushy, frosted side-whiskers gave him a jolly, Dickensian pomposity. The birds ignored him. The red squirrels moved in to retake the corn pile, but retreated when he lifted his head. The chipmunks stared with astonished eyes, hopped near in little jerks, almost touched him, and leaped away. He considered them too small to merit his attention.

He made forays through the wild greenery within a hundred-foot radius of the cabin. He climbed the fence awkwardly to finish off the pea vines. He shinnied up posts to get at the damaged bean plants that we had not bothered to protect, and clung precariously with hind feet and one arm, while he waved the other until he captured a distant leaf. He eyed the rest of the beans longingly from outside their new fence. He seemed to think the carrot bed too dangerous for daylight feeding and demolished the leaves in midnight snacks. He waited patiently until my bed of violas had a nice crop of blooms before he ate them. He accepted crackers from me, held them daintily to his mouth with one paw, then gobbled so fast that crumbs dribbled down his vest.

He grew fatter and fatter. In early September he tried to climb a post after a last bean leaf, but gave it up and slid breathlessly to the ground. His belly dragged as he waddled to the corn pile, and he lay down every few feet to rest as he went back to his burrow. He settled to sleep on September eleventh, when the days were still warm, maybe because he was so fat that he could no longer comfortably manage to pull himself to the food.

Back of the garden, beyond the drainage ditch, is a stony bank. The ditch and bank are covered by a pile of branches that are the remains of cedar fed to deer, and ferns and raspberries make this into a green tangle. Under this excellent cover, between two large stones, is Gregory's main doorway. The pile of dirt usually found outside woodchuck burrows and on which they stand to look for danger before venturing forth was washed away by the torrential rains that accompanied Hurricane Carla's passing. But Gregory's entrance is so well protected that he does not need an observation mound, and we have never heard the shrill danger signal that gives woodchuck the name "whistle pig."

Now Groundhog Day is five days past. Gregory's bank and brush pile are blanketed by three feet of snow. Secure in his burrow, he is curled up, living on his layer of converted corn and flowers. His body is chilly, his heartbeat and breathing are slow. The North Woods has a long stretch of winter after February second, so why should a groundhog wake up?

We look forward to his waking, especially as it has occurred to us that Gregory might turn out to be Grace. A yardful of baby groundhogs would be hilarious—and it is *much* harder to cultivate a garden than to buy canned vegetables.

In April, when the weather is warming but snow is still on the ground, a line of small tracks emerges from the brushpile. They circle and zigzag, then turn back to disappear under the tangle. The chipmunks that live in the bank have stretched and come out to test the night air after their six-month retirement.

We have two species, the least chipmunk, a western rodent, and its cousin, the eastern. Only in Upper Michigan, northern Wisconsin and Minnesota, and a small northeastern corner of North Dakota, do their ranges overlap.

The least chipmunk is slim, only three to four and a half inches from nose to rump, with a flat, thinly haired tail that is about the same length, but looks longer. Its fur is grayish-tan, marked by five black stripes down the back all the way to the tail base, with grayish-white between the pairs that lie along the sides. Its face is strongly striped in black and white.

The eastern chipmunk is larger and chubbier, with a tail not as long as the rest of it. Its face is more strongly striped than that of the least chippy. It also has five black stripes down its back, but they end at the top of the rump, and the white marking between the side pairs is bright and lenticular. This chippy is brushed with deep red on its rump and along its back and, at close hand, is a study in beautiful coloring.

The least chippies do very well in our yard. Ade has built several stone piles, each of which protects the burrow of one or two families. Increasing numbers tell us that they are biologically successful—a mother may bear two litters of three to seven each in a summer, although the young are so large when we first see them that we can tell them from the parents with certainty only when they are side by side. Last fall a new home was burrowed under the stones of our "groundhog fence," and its occupants sat for hours on the fence posts in the sun, grooming their fur or quietly chatting in small chirps like dripping water. Usually these chippies are extremely active. It is startling to see so small a mammal running straight to the top of a pine or spruce that rises over a hundred feet.

The eastern chipmunks seldom climb anything taller than a honeysuckle bush, and they move more slowly than the others. I think this lesser speed accounts for their smaller numbers in the yard, although it is thought that they produce only one litter of three to five a year. The surroundings and food are just as good for them as for the least chippies, but their enemies—the weasels and fishers and foxes, the hawks and owls—may catch them more easily.

When our chipmunks first come out of their burrows in spring, their natural food is scarce. They head straight for the corn, stuff their cheek pouches, and lay in a quick supply against late snowstorms. They are fond of graham crackers, and dash up to beg so quickly and so near that I have had to fall sideways to avoid crushing one with my foot.

They look eagerly for the first spring greens. They break off the new everlasting leaves and hold a leaf section in both paws, nibbling it like a melon slice. The next delicacy is dandelion seed heads, which they nip off after laboriously pulling the tall stems down to their level. If I want to keep them from discovering my alyssum sprouts, I pick the dandelion heads and leave them in convenient piles near the burrows. Then come wild strawberries, which the chippies search out amid their grass forests. As the raspberries ripen, the little mammals scurry up and down the prickly stems, eating the fruit and licking their pink-stained fingers. Later they cling to the climbing false buckwheat that I have trained up the cabin-side, reaching and slipping, catching a lower branch, eating the black pyramidal seeds. Insects are welcome, too, especially carpenter ants and various easily obtained pupae. Chipmunks are often condemned for eating birds' eggs, but our birds show no alarm in their presence.

Some years ago we had a black, three-foot, pet rooster, with a strong tendency to resent intrusion in his domain. But the chippies came and went, standing hip-deep in his corn pan and carrying his food away by pouchfuls. He cocked his head and made henlike clucking sounds as they scampered between his spurred legs. Ade insists that he considered them his pets.

Last summer I planted some castor beans, a silly thing to do when one considers the difference in climate between the beans' native Madagascar and our Minnesota border country. By luck and labor one plant thrived, standing like a small tree with a bright-red stem and two-foot-wide bronze leaves. One of the least chipmunks liked to relax at its base

on the soft moss with which I had mulched the soil, his tailtip twitching gently, his striped body stretched in the shade, a small tiger under a jungle tree. With all of his siblings, he is drowsing now, waking occasionally to eat a snack of his laboriously stored corn. And in some other buried place the eastern chipmunks are lying in deeper sleep. Before I have finished this book, they will all have come out into the sun—and I will plant some castor beans. My miniature tiger, if he remembers, may again want to lie in the shade of his own small tree.

The winter sleep of the black bears differs little from ordinary sleep except in its length. The temperature remains high and the deep, slow breathing is like that of a soundly sleeping human. A friend of ours came on a big male, snoozing in a cavity under the roots of a fallen pine. He poked with a stick until the bear awoke, slapped the offending wood away, and settled drowsily back. The investigator, ignoring the hint, poked some more. The bear snapped wide awake and started clawing through the snow and debris that hampered his movements. Our friend took off at such speed that (he says) the snow melted under him.

In January, mother bear dozes through the birth of her cubs. Beside her two- to four-hundred-pound bulk, these nine-inch, half-pound mites look incongruous. They are blind, but smell the way to her milk; they are hairless, but keep warm in her fur. Usually they are twins, but sometimes triplets or quadruplets, and a mother followed by five cubs of the same size was seen by Superintendent J. A. Wood, of Prince Albert Park, Saskatchewan. For two springs on our shore, an old bear has been seen leading, in single file, her five-month-old twins and her two yearlings. The yearlings may have denned up with mother when the snow came in November, or had their own quarters nearby, from which they joined her at wake-up time in April.

The bears here are black with brown faces, as are all eastern black bears. In the West, litters may contain the cinnamon phase that may be light to dark or reddish brown. In southeastern Alaska a beautiful bluish-gray cub is sometimes born and, in British Columbia, there are occasional almost-white albinos. (In May, 1963, a cinnamon cub appeared on our shore. Unfortunately, someone killed this animal, rare and perhaps unprecedented here, and discarded its carcass on a nearby dump.)

Wild black bears scoot for the tall timber at the sight of man, but half-tame bears gather at garbage dumps and forage around cabins where the smell of food is enticing. People remember that the grizzly and the big brown bear are meat eaters but forget that the black bear also is a carnivore, although its diet embraces about everything else from flour to blueberries. Its comical and cuddly appearance has largely overcome prejudice, but has caused admirers to lose reasonable caution in its presence. A local newspaper once published an urgent warning, pointing out the danger of dangling food on fish poles to bears feeding at the dump—and jerking it back when a bear reached for it! One might remember that meetings with black bears can produce highly varied effects.

There was the day I stepped out the door just in time to feel a rush of air and sense a black bulk speeding past. The bear said *"Whufff!"* and kept on going. I gargled incoherently and fled back inside.

Several outbuildings stand fifty feet from the house with a yard in front that, at that time, contained chickens. The bear stopped in a rustic arbor beside the buildings. He shifted from side to side, claws clicking on the flagstones of the walk, head swinging back and forth, apparently undecided whether or not it would be prudent to approach the house. One pullet, of the clucking row that pressed against their wire enclosure, decided this was cause for excitement. Her squawking and flapping set off the others. The bear wheeled away from this unfamiliar racket and loped into the surrounding big woods.

Reactions to this anecdote, both from tourists and local people, range from, "You might have been killed!" to "Pooh! Bears aren't dangerous." Both of these conflicting statements contain truth. The bear, according to a woodsman who saw it shortly after I did, weighed five hundred pounds. This is very large; black bears usually weigh from one hundred pounds, for a yearling, up to four hundred pounds. A mere collision with anything so heavy, moving in the deceptively awkward-looking lope which reaches speeds up to twenty-five miles per hour, would have damaged me severely. Had we collided, the frightened animal probably would have struck at me instinctively. Had that occurred, no doubt I would not now be telling the tale. On the other hand, had I looked where I was going, the bear, headed somewhere in a hurry, would have dashed by, ignoring me completely.

Last spring a sleek black mother with two matching cubs approached the log cabin. While the cubs sniffed around a stump, mother put a paw on the sill of the open window out of which I was looking and stared in at me. We could have rubbed noses. I did not move, kept perfectly calm, and looked back with equal curiosity. She made a soft sound that recalled her babies, now wandering off in separate directions. The three of them stopped to sniff the Johnny-jump-ups in a flower bed and, in the way of all bears, ambled away, intent on their own affairs.

This pleasant encounter had elements of real danger. Had I moved, had I been frightened or tense, mother bear's keen senses would have signaled, "Danger!" She could have broken my neck with one swipe of her paw, and would have if she had felt that I menaced her cubs.

Once I looked through the glass panel of the kitchen door into the wondering eyes of a very little cub, standing spraddled-legged on the step and sniffing rapturously at some odor from within. This done to his satisfaction, he walked away a few steps and stood up, waving his small "arms" to keep his balance. His round head seemed too large for his thin neck to support, and his belly showed a definite little pot. Nosing in a hollow we were filling with trash, he found an accidentally discarded can of milk. He rolled this to a convenient flat stone, riddled it with tooth holes, and happily lapped up the contents. (He did not, as someone once suggested, read the label! Apparently well-educated in selecting garbage, he may have been attracted to the can by its weight.)

Surely no young animal is more amusing and appealing than a black bear cub. It would have been fun to make friends with this youngster, but he might have had a watchful mother out of sight in the brush. It required determination not to put out food for him, but I remembered that within a year he would weigh a hundred pounds, and keep right on growing—while he still expected food at his baby feeding station. His strength and appetite would keep pace with his size, and he might decide to push a door off its hinges and look inside for lunch.

The big fellows follow their ancient trail along the hill above the house. They amble down the road or leave footprints in the soft garden soil. Ade and I remember that these mighty wilderness citizens are naturally amiable, but, if frightened, injured, or angered, can be ferocious. We remind ourselves that, in spite of their engaging appearance and manners, animals strong enough to bite through inch-thick boards and to rip open logs with their claws are not safe pets. We enjoy black bears at a distance.

The Wanderers

ON A BITTER DAY near the end of winter, I saw the backs of three brown-striped, brown-winged, chickadee-size birds that were eating rolled oats on the bench under the cedar tree. I caught a flash of red on a head and decided that the birds were ruby-crowned kinglets. I went into a dither. I was quite sure they had arrived ahead of the season. There would be no insects for weeks. Without these for food and warmth, the birds would surely starve. I was envisioning their pitiful, frozen bodies on the snow when I took another look and woke up. I had fallen into two of the mental traps that I warn other people of.

In the first place, I had tried to identify the birds without really looking at them. Kinglets are greenish-gray, not brown, are even smaller than chickadees, and the red headpatch of the ruby-crowned male usually is concealed, except when he is displaying. The golden-crowned kinglet, whose head has a conspicuous yellow or orange stripe bordered by black lines, winters in woods like this, associating with the chickadees, but I have seen none except at migration time. The birds on the bench were common redpolls. All had bright-red patches above the beak and well-marked black chin patches; one showed the male's beautiful pink breast.

In the second place, I had forgotten that not all strange birds seen in the spring are returning from warmer areas. The previous fall, these redpolls had moved south into the northern states from their far-north nesting grounds. They were now returning north to lay their half-dozen eggs in grassy nests located in bushes or small trees. While they fed with us, they might shelter in the thickets of alder or young birch nearby. Like other seed eaters, they can winter in climates where insect-eating birds, such as warblers, would perish in a day or two. Plump and cheery, the redpolls left the bench and clung to weed stalks released from the snow by an earlier thaw. As long as there were seeds to be found, above the snow or in the trees, they would go on their way, perhaps to meet a northbound flock containing their cousins, the hoary redpolls, whose trek would extend to the very limit of Arctic land.

I have not seen any more redpolls here, but they may have passed by in the treetops with the flocking pine siskins. The siskin looks like a slender redpoll, without the chin and head patches, and with strong brown striping on the back and breast, and yellow patches on wing and tail. They wander without routine, according to whims known only to themselves, nesting here or there in conifers as the terrain strikes their fancy. I try to imagine them, tending three or four young in a saucer-shaped nest hidden in our tall evergreens, but the birds are so much on the move when I see them that this is hard to do.

A sizzling, buzzing twitter signals the approach of a pine siskin flock. The birds appear in darting masses, like leaves tossing high on the wind. They swirl and gather, then settle into the tops of the tallest trees. Twigs bend with their landings, snap upward as the birds change perches. The shelflike branches of the old white pines seem covered with misplaced, animated cones and the tops of the dignified black

spruces dance. The pale green of the birches is darkened with brown patches of birds in summer, and their bare limbs are clothed transiently with feathers in winter. Now and then a few siskins may pause in the feeding yard; one spent several midsummer hours in a neighbor's work building, resting on a beam. Goldfinches accompany the siskins in warm weather, and the bright yellow of the male goldfinch's breeding plumage flashes as the flock flutters and wheels across the sky.

White-winged and red crossbills travel as erratically as the siskins, nesting where they will. They may drift into our trees with the siskins or they may come alone at any time of the year.

Five years ago a monster of a spruce began to lean so that its cones fell on a path where it was easy to collect them. We scattered a bushel of them on the snow the following January. The cones attracted a small flock of white-winged crossbills, the males a brilliant, deep rose-pink, with two strong white bars on each black wing, and the females a modest olive color, with the same striking wings. The birds reached into the cones with the crossed, curved mandibles of their beaks, pried open or nipped off the bracts, making little clicking sounds, and extracted the seeds with ease and dispatch. Then they were off, chattering and warbling, leaving behind a pile of shattered cones and the memory of color and life against the snow.

The male red crossbill is a dark, rich red; the female, olive. Both have black, unbarred wings. A pair of these nested high in a spruce between the cabins the summer after we had seen the white-winged flock. They warbled sweetly for us during the spring and came into the yard for sunflower seeds several times with their two young, whose brown striping made them look like oversize siskins. In a late summer flock with the siskins, only their larger size and lack of yellow patches

set these youngsters apart. The crossbills cling in a parrot-like manner to branches when they feed; I have been able to identify some juvenile crossbills in a siskin flock by this habit. Some of these may have been the offspring of the white-winged birds that fed on our spruce cones during their very early mating season—January and February, when the snow and cold are at their deepest here. The juvenile white-wings have two characteristic white wingbars that the young red crossbills lack, but my view was not clear enough to see this detail.

Last fall I had just finished watering my sunflowers, which had stretched to ten and eleven feet as they reached for light, and was admiring the brightness of their big, yellow crowns in the dusk, when I heard what sounded like a confusion of whistles and the tip-tap-tip of miniature hammers. I was highly entertained by a vision of gnomes, whistling as they forged a leprechaun's gold.

The sound came nearer through the trees and I saw birds, moving and settling in the shadowy peaks of the evergreens. With what seemed a contrived perverseness, they avoided the high streaks of light from behind the hills where the sun was setting, so that I could not get a glimpse of either size or color. Then I heard a sound of wings behind me. A male evening grosbeak was perched on a fence post not two feet away, examining both me and the sunflowers with interest.

His forehead, beneath his ruffled black crown, was as yellow as the flower petals, and his heavy ivory beak contrasted exquisitely with the bronze of his head and neck, which blended and paled into deep gold on his back and underside. He flipped his black tail and bent his head to preen the feathers of his right wing, spreading the black primaries and white secondaries like some exotic fan. Hav-

ing identified the flock for me in the best way possible, he whistled and returned to the treetops.

I had only a small quantity of sunflower seeds—the growing plants were immature—but I spread them on a bench early the next morning. The grosbeaks, about thirty in all, made short work of them. The females, cracking the hulls with their pale, powerful beaks, were no less beautiful to watch than the males, although less spectacular, with their lustrous gray feathers, yellow shadings, and black-and-white wing and tail accents. The chickadees were highly annoyed at these big birds that were eating all their seeds and dived in and out, snatching breakfast from almost under the feet of the grosbeaks.

To feed even this small flock through the winter would have taken a large quantity of seeds, which I could not get because our summer-only freight deliveries had already stopped. They might have adapted to other foods, but this usually takes time. We fed gray jays for four years before they discovered that cracked corn is edible. So the grosbeaks moved on, possibly to winter near one of the feeding stations maintained in the yards and gardens of the villages along the North Shore of Lake Superior.

The comings and goings of these birds are clothed in mystery. They nest near the top of evergreens in the forests of the north and west, rear up to four young, and apparently winter wherever there is suitable food. Banded birds have been observed to travel east and west, between Minnesota and New Hampshire. They were rarely seen in the Cleveland area before 1945, but, with rising interest in bird identification and an increase in feeding stations, flocks of more than a hundred, accompanied by American, or common, goldfinches in dull winter plumage, visited feeders there in the winter of 1961–1962.

The absence of balsam fir and spruce cones, caused by the budworm infestation that rose in Canada and rapidly spread southward about ten years ago, may have sent grosbeaks south in search of food. The almost complete failure of the pine- and cedar-cone crop and the destruction of vast forest acreage by fire, both related to the severe drought of 1961, may have increased the movement.

A flock of evening grosbeaks is feeding this summer at a lodge only ten miles away. When the lodge closes and its feeders are empty, I will have sunflower seeds waiting and may be lucky enough to attract this flock. These handsome birds are amazingly fearless in approaching houses, so anyone who lives near wooded areas that supply shelter may have the good fortune to attract some of them. Box elder and hackberry fruits are among their favorite foods. If assistance can extend their range and increase their numbers, the expense and effort of feeding will be amply repaid.

The pine grosbeaks nest as far north as they can find evergreen trees and travel irregularly through the winter months, looking for the pines and spruces that mean home, no matter how temporary, to them.

We see them singly, or by twos and threes, in winter. They could pass unnoticed if it were not for their odd tracks—the prints of the feet, then two short, straight, dragged lines to the next footprints, moving this way and that in uneven zigzags. When we see the tracks, Ade and I spell each other in watching for the birds' return.

They are as handsome as the evening grosbeaks and as large as robins, with notched black tails, white-barred black wings, and black beaks with a swollen look, caused by the strong curve of the upper mandible. The male has rose-red on breast, head, back, and rump; the female is a soft gray in these areas, with yellow crown and rump feathers. The immature birds resemble mother, except that males have reddish touches on crown and rump. There are three or four to a brood if all eggs hatch and nestlings survive. Sometimes these grosbeaks are so unafraid that, on their first day with us, we can step past them while they feed, and I have heard of instances where parents brought very young fledglings to feeders.

In 1956, deep snows were followed by strong winds that brought down clouds of seeds from the previous summer's big crop of cones. Pine grosbeaks, in flocks of from a dozen to fifty, moved back and forth over the snow during late January and February, resting in the trees during windy periods, then returning to eat the newly fallen seeds. Our efforts to bring them into the feeding circle failed, although they took cedar seeds from the snow just outside the door. Perhaps they were driven instinctively to move on. Because the few individuals that have fed here have been so ready to endure human company, I think that feeding stations may in time acquaint more people with these birds, but only where there are evergreens, native or cultivated, for the pine grosbeak was rightly named. It would not feel at home among the bare winter branches of deciduous trees.

Occasionally, when the snow is not too deep to bury the seed heads, or when a thaw has released some of those buried, we see a few snow buntings in the open space around the log cabin. Chubby and lively, they fly against the weed tops to knock the seeds loose. Against the whiteness of the snow and of their underfeathers, the yellow markings of head, chest, and rump and the dark patterns of back, wings, and tail are very attractive. Only when they fly do their white underparts show plainly, and flocks that wheel over the open lands must look like living snowstorms. Thence comes the snow bunting's common name, snowbird, which is also applied to the junco, who picks seeds from the snow of spring and fall and may winter in snowy locations.

It is not strange that snow buntings seldom come our way, for they have no interest in trees. They breed in the open

area, from the subarctic tundra to within at least seven degrees, or less than five hundred miles, of the North Pole, a
harsh land indeed for the hatching of their four to eight
eggs and the rearing of their young. They are friendly oppor-
tunists when in their northern range, often visiting camps
and villages, where they are called arctic house sparrows. In
their wintering range, south from Unalaska Island and north
ern Quebec into the United States, they frequent fields and
prairies, beaches and bare mountainsides. Sometimes they
pass our small open space without our seeing them, leaving
only small foot and wing marks near the weed tops, and
rounded hollows under clumps of tall grass where they have
spent the night. And once we found some scattered feathers
and the track of a fox, silent reminders of a bunting that had
settled too close to the brush and slept too soundly.

All of these erratic travelers are of the finch family, and
one of their greatest charms is the capriciousness with which
they appear. Against the stable background of the birds that
are always with us and those that come so regularly that
their absence is cause for concern, these vagabonds move
with the beauty of a double rainbow, or a red aurora, or any
other rare and unexpected thing. So it was with the pair of
purple finches that came in the spring of 1960.

They arrived with the migrating sparrows and looked
much like sparrows, except that the male was touched with
a fuschia-red more like light than color. In shadow, he
looked merely ruddy brown; in slanting sun, his head and
rump shone like fluorescent rubies. They lingered several
days in the yard, the little striped female hopping sub-
missively at the heels of the regally colored male, as they
stuffed themselves with seeds. Then they vanished.

A few days later, a most beautiful song came from the
edge of the forest, an unrestrained and jubilant warbling
that stirred memories of the unshadowed happiness that be-

longs to childhood. The finches were nesting in a balsam fir twenty feet from the cabin.

Through most of the summer that wonderful, heart-warming song rang out, the male finch's reaffirmation that this was his place in the woods. In August we saw both parents with two young in the branches of a bushy, ten-foot white spruce. As they lay four to five eggs, their percentage of success matched that of our resident blue and gray jays that lay four, and three to six eggs, respectively, and produce two young on an average.

Perhaps the finches will return, but perhaps not—for they are of the wanderers, and are to be enjoyed to the fullest on those occasions when they bring their special beauties within the range of human senses.

The Air Hunters

THE WHISTLING ROAR of a jet plane broke the stillness of an autumn afternoon and I looked up, to watch a silver spangle draw its white ice trail across the clear sky. Far below, a bald eagle crossed the spreading streak at a right angle. Its white head and tail stood out against the deep blue and its dark body seemed one with the great wings that steadily raised and lowered, driving powerfully ahead. Behind came a second eagle, wingbeats timed with those of its mate, and, following, their dark-feathered, immature young one. As the contrail drifted away in the upper air, the migrating family flew out of sight beyond the southern hills.

In the past, the bald eagle's six- to eight-foot wingspread and its carnivorous feeding fostered ridiculous tales of its stealing half-grown sheep and even children. This, combined with the fact that so large a bird makes a splendid target, encouraged widespread shooting. Many others were accidentally destroyed when they tried to steal fish or meat bait from spring traps. Although the bald eagle was chosen as the national emblem of the United States in 1782, it was not given protection until 1940 and, in Alaska, 1952.

The apparent decline of the species has led to the Continental Bald Eagle Project of the National Audubon Society. The Audubon staff have been assisted by Federal and state conservation agencies, private conservation organizations, and many interested individuals. At the end of two years of a five-year study much has been learned about eagle numbers and movements. A nationwide count in January, 1962, showed 3,807 bald eagles in the United States (exclusive of Alaska whose population will be studied later). Fifty-seven per cent of these wintering eagles were seen in the Midwest, mostly along the Mississippi River from southern Illinois to Minnesota. Dams in this area help to keep the water open in winter and the associated generating plants supply injured fish as food for the eagles. Florida reported 529 birds (fourteen per cent); the majority were nesting. The Northwest accounted for ten per cent and the mid-Atlantic states for six per cent of the total. The remaining thirteen per cent were scattered throughout other areas of the United States. Only twenty-four per cent of all eagles counted were reported to be immature birds, but not enough is known at present to indicate whether this is abnormal or not. (Anyone wishing to aid in the "eagle count" or to report eagle nests may obtain information from: National Audubon Society, Box 231, Tavernier, Florida.)

The bald eagles that I saw migrating nest on the Canadian shore, probably at the top of one of the tall pine stubs left after a fire in 1936. The young bird might have been from a pair of eggs laid the previous year, because it may take three years for an eagle to grow its mature plumage.

Although the bald eagle's beak is designed for tearing flesh, its feet are too weak to permit it to lift much weight. It picks up dead fish and crippled small mammals near water; it bullies the osprey into dropping freshly caught fish which the eagle then devours; and sometimes it catches waterfowl. In the winter I have seen bald eagles driving ravens away from wolf-killed deer carcasses on the lake ice. Scavenging may not be an elegant way of making a living, but it is harmless and useful.

Any bald eagle in flight is something to remember and the sight of a youngster brings a flare of hope for the future, but a close-up view of an unconfined adult is awesome.

In January, 1958, an enormous, wide-winged shadow slid across the snow in the clearing as an eagle glided down to light on a small storage shed twelve feet from the log cabin. It gripped the edge of the roof with its taloned yellow feet and sat, brown body tense, brown wings half-lifted, tail folded into a white strip, and snowy head thrust forward. The heavy yellow beak, with a powerful hook on the upper mandible, was almost as long as the head. Feathers grew low and straight across, like a scowling brow, above the fierce, golden eye. The mighty bird perched for some minutes, looking right and left, then sprang out and up on huge, lifting wings. It rose in a great spiral. Gradually it diminished in size until it seemed black and headless, then only a dot. Then, so high that it was a pinpoint, it disappeared in the sea of light.

I thought of the men who selected the high-flying eagle as our national emblem, and of later critics who suggested that the wild turkey would have been a better choice. I wondered if these critics had considered the present status of the two birds. The bald eagles, though not so numerous as they once were, still fly free. Wild turkeys have largely lost their identity through interbreeding with domestic stocks and their small numbers are recovering in some areas only because they are being bred and stocked by men; their destiny seems to be the oven. My vote goes to the eagle.

The shrikes are present here in two hard-to-distinguish species. The loggerhead shrike breeds throughout temperate North America, laying four to eight eggs in a nest located in a bush or tree, usually in farming country. It appears here occasionally in late summer or when migrating. The northern shrike lays the same number of eggs, but breeds in openings of the farthest limits of the boreal forests. Individuals winter here, but they are not common.

Although they have the hooked beaks of flesh eaters, shrikes have weak feet and no talons. The loggerhead feeds largely on insects, which do not have to be gripped firmly while being eaten. The northern shrike consumes small birds and mammals, snakes, and insects, with a frog now and then. It may grip the prey with one foot, the leg lying across its perch and the food hanging from the foot, or it may impale its victim on a sharp thorn or twig, as an aid to holding it while tearing it with the beak. A thorn bush, decorated with this bird's accumulated food, is a gruesome sight, and the custom has given the northern shrike the unpleasant name of butcherbird.

A friend once told me that she had seen a gray jay attack a chickadee that had been stunned when it flew against a windowpane. The attacker was surely a northern shrike, whose gray and black and white markings are somewhat like those of the gray jay. The jay, although a carrion eater, never attacks a living bird or mammal, even though it be lying motionless.

We have never seen a northern shrike in our feeding yard. I have little doubt that such a visitor would be promptly driven away by our cooperative battalion of blue and gray jays, with embattled chickadees bringing up the rear. But I am not sure, and I do not like to think of one of those fat chickadees, today sitting on my finger and picking crumbs from my palm, tomorrow perhaps impaled on a twig for a shrike's dinner.

On the other hand, I am downright fond of owls, although they do not kill gently. We are well supplied with them, particularly with barred owls, which find many tree cavities and old squirrels' nests that may be used for the rearing of their two or three young. Mother owl lays her eggs on the accumulated rubbish in the nest and does little by way of improvement. These owls stay for years at one location, and the nesting cavity gradually becomes lined with downy feathers shed by the young as they grow adult plumage.

The barred owls are so much a part of our background that we wonder what has happened when we do not hear them for a week or two. *"Hoo-waaah!"* booms from the dark woods. *"Hoo-hoo-hoohoo, hoo-hoo-hoohoo-wah!"* comes the answer in a melodious treble. Sometimes these owls gather near the cabin and hold long conversations. They chuckle and hiss, whisper and grunt, cackle and hoot, in such a variety of sound that hearing them is like listening to a newscast in an unfamiliar language; one strains his ears on the chance of hearing a familiar word. One of these owls had a favorite hooting perch in the big pine west of the log cabin. One night, in the midst of a series of gentle *hoo-hoo's,* it stopped, presumably took a deep breath, and screamed—a long, horrifying shriek that ended in a bubbling gurgle as of something with its throat cut. It twice repeated its ghastly cadenza. Then, as though satisfied with its performance, went back to its pleasant hoo-hooing.

Barred owls see very well in the light and do much of their hunting by day. There was an October afternoon when an eighteen-inch individual settled on a branch overlooking the feeding yard and prepared to select its dinner from the small visitors. Its brown eyes, encircled by pale feather disks, and its sweeping feather moustache gave it a look of amazed good-nature. The cross-barring of its breast extended upward to surround its head, which appeared as a large feather ball. As it leaned forward to watch the ground, its brown-and-white streaked belly feathers covered its downy white feet. The brown and gray and white patterning of its back blended perfectly into the tangle beyond.

To my surprise, neither the other birds nor the mammals paid any attention to the owl. It looked around in a leisurely fashion, then concentrated on the earth beneath, where something tasty might run out from under the woodpile. Then I noticed that, although most of the birds were still feeding, all the blue jays but one had withdrawn to high branches. This remaining jay flew from low branch to low branch, always keeping the owl in view. Abruptly the hunter launched itself toward an unwary young squirrel. As easily as though it had been rehearsed, the blue jay flew in an arc that just intersected the owl's flight and spoiled its aim. The squirrel hopped away, the jay preened on a branch, and the disgruntled owl stood flat-footed on the ground. It whirled its head around in that way which almost convinces one that owl heads are screwed on, looked at the jay as though thinking the woods would be a pleasanter place without such interfering creatures, and flew off to a less-policed hunting ground.

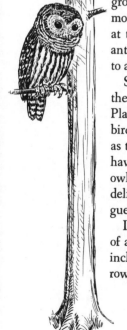

Several people have asked me if this move was planned by the jay and, if so, why. There are no satisfactory answers. Planning and reasoning, in the human sense, do not apply to birds. The group behavior of the blue jays was uncommon, as they usually gather in a noisy flock and pester an owl or hawk until it retreats. The single jay's interception of the owl's flight was so perfectly timed that it *appeared* to be deliberate. Beyond that, anything I might say would be guesswork.

In almost ridiculous contrast to the twenty-four inches of a large female barred owl is the seven- to eight-and-a-half-inch saw-whet owl, about the size of a white-throated sparrow. This woods has many old woodpecker holes where the little owls may lay their three to seven eggs. We knew they were present because on February and March nights we heard what sounded like the filing of a saw when no one was near who might actually be doing so. Although these owls, especially in remote areas, are quite tame, we had little hope of seeing them because they are largely nocturnal, but a suet feeder, forgotten at night, attracted one.

I was looking at the moonlight that turned falling frost into blue and white sparks when I saw the little owl, standing on a horizontal limb just above the suet cage. It considered the suet gravely, turning its head from side to side, then flew to perch on top of the cage. It worked patiently with its beak but could not extract the food through the wire mesh. Nothing daunted, it flew back to the branch, where it paced one way and then the other, bobbing its head and giving the impression of having its wings folded like arms behind its back. After due consideration, it tried to cling to the tree trunk and thus attack the food, but had to give up and go back to branch-pacing. I opened the door a crack and tossed a small chunk of suet to the base of the tree. The owl halted with half-lifted wings, then sidled to the trunk and leaned so far forward to watch the door that I thought it would tumble off its perch. Satisfied, it drifted to the ground and circled the suet on foot, beak open and ready to return any attack. Then it flew up, circled once, and pounced successfully on its "prey."

Since owls are thought to have only a rudimentary sense of smell, I wondered how it discovered the suet, which was surely unfamiliar and in no way resembled a mouse or other ordinary owl food. We have often seen birds perching and watching the regulars feed, after which the newcomers ventured to approach. Perhaps the saw-whet had done the same during the daylight hours and had come in to see what the attraction was.

It visited regularly for several weeks, and, on a night bright enough for good owl-watching, I set a large ham bone, with fragments of meat and rich marrow, on the shelf by the door. The little owl soon arrived, clutched the bone with one foot, and tore the meat away in strips. After it removed each morsel, it leaned far back to peer at the bone in the strained manner of one who needs, but will not break down and buy, reading glasses. (This is not so far-fetched as it sounds. Although owls have unusually sharp distance vision, they cannot adjust their eyes to see well at close range.) It was undergoing one of these farsighted stretches when it saw me inside the glass of the door, not two feet away. It stretched a wing protectively in front of the precious bone and puffed up to twice its usual size, snapping its beak and making a hissing squeak. The miniature mouser looked very fierce indeed; I "fled" from the door and in the morning the bone was cleaned.

Although I have not seen any screech owls here, I have infrequently heard ululating calls that might be theirs, and that probably were made by transients, for these owls prefer more open nesting sites. The idea that they screech may have originated when the fearsome shriek of the barred owl, or other creature, was heard by someone who saw screech owls in the vicinity.

A pair of these owls nested in the basement of my childhood home. They were gray (although there is also a rufous phase), and could not be mistaken for any other bird, because screech owls are the only small owls with "horns." My father, appreciating their value as mousers, left a window open for them and they reared three young. The parent birds protected the furnace room very effectively against at-

tempts to pry into their domestic affairs. Sometimes they called in the night and the soft notes crept up in ghostly echoes through the hot-air ducts. When the winter coal was delivered, the owl family moved out and I last saw them sitting in a row on the telephone wire that was strung on poles along the street curbing.

The boreal, or Richardson's, owl, which rears four to six young in coniferous forest like this, but farther north, visits us irregularly in very cold weather. One of them once spent an afternoon hour on a branch near the cabin. It looked much like an enlarged saw-whet, wearing big, black, horn-rimmed glasses. Its forehead was spotted whereas that of the saw-whet is striped vertically, but this distinction may be obscured by windblown or otherwise ruffled feathers. It seemed restless and displeased with the other birds and, when the blue jays began to pester it, went off without delay.

Another winter visitor, and one of the most beautiful, is the snowy owl that comes from the Arctic in search of food about every fourth year. These owls nest on hummocks in the open northern barrens. The nest is a mere hollow, often surrounded by snow. The mother begins incubating the first egg as soon as it is laid and, at irregular intervals, lays more until the clutch reaches from three to thirteen. The eggs hatch in the order laid and the unhatched later ones are often partially covered by downy nestlings while the mother procures food for the eldest of the brood, which may be fully fledged when the last egg hatches.

The females and immature birds are heavily barred and patterned and are not so easily recognized as the white, lightly barred male. We had a fine view of a male, sitting on a fence post by the road to town as we were driving in for

supplies four months after we moved here. Ade stopped and we exchanged curious glances with the owl, which was side-lighted and had squinted the eye on the brighter side so that it wore a most dissolute and tipsy expression. We could even see what appeared to be short horns, almost hidden by other head feathers. It did not seem large until it spread wings like white triangles and sailed away, legs swept back and swaying in a negligent manner.

Ade came in one night last December, looking shaken, and asked me if I thought we just might have ghosts around the place. Something white and silent had flown past his face, almost brushing his nose. Two nights later the ghost flew up from a stump as I opened the door, a burst of white that looked like a clump of snow, exploding upward and disintegrating as it reached the level of the lower branches. An owl it had to be, because of its soundless flight, and probably a snowy, although I sometimes think wistfully that it *might* have been the rare pale arctic horned owl.

We have been told that the more usual brown great horned owl is resident here, but have never heard its booming. This two-foot fellow, with tall feather horns, white chin ruff, and round golden eyes looks like no other bird. However, several times we have heard the hooting of the barred owl misidentified as that of the great horned. There is a question, too, as to whether the screams attributed to the great horned owl are not those of the barred owl.

The great horned usually lays only two eggs in thick woods. It is the largest common owl in North America. Its appetite for rodents makes it beneficial, except when food shortage forces it to raid poultry yards. There are a number of races of horned owls and, cumulatively, they nest over the Americas from the Arctic Archipelago to the Strait of Magellan, excepting the Caribbean Islands.

I dream of seeing the great gray owl that nests in the boreal forests as far north as the tree limit, as far south as the northern rim of the United States. Its nests, containing three eggs, have been found not far from our location, and the owl occasionally migrates somewhat south and east of here in winter. I know what it looks like: a three-foot, smoky version of the barred owl, with yellow eyes, breast and belly striped and faintly barred from throat to tail, and a six-foot wingspread. I also know that if I am ever granted a sight of this bird, no photograph or painting or stretch of my imagination will have attained its mysterious beauty and dark dignity.

The best reminder I have ever had of keeping an unprejudiced mind toward wild things came from a lady visitor. She was feeding young gray jays when I called her attention to a goshawk, maneuvering its forty-inch wingspread through the brush with not-quite-credible speed and dexterity. "Goodness!" exclaimed my visitor. "You ought to shoot that hawk. You wouldn't want anything around here that would kill a *bird!*"

The goshawk is one of the accipiters, along with the sharp-shinned and Cooper's hawks. These birds have a distinctive shape when flying, with a long slender tail and stubby wings that aid their headlong flight through the forest. They do feed on other birds, but not exclusively. All three of them are present off and on in our yard in summer, and the goshawks nest here, but our bird population remains stable. If the hawks were not here, the other birds would increase, but this would lead to starvation for some of the resident birds if we were not here to feed the overpopulation in winter.

All three accipiters nest in thick woods, usually laying from three to four eggs. The goshawk finds either coniferous or mixed growth suitable, and the very large nest on our property is placed some sixty feet up in the crotch of an old birch. Cooper's hawk prefers coniferous forest, and the sharp-shin likes coniferous edges. Probably those individuals of the last two species that we see are rearing their young in nearby seclusion.

The twenty-one- to twenty-five-inch goshawk is the largest of the three and, when mature, is gray, with a lightly barred gray-and-white breast and a strong white line above the eye. A young one once took up a waiting position atop a stone pile which protects the entrance to a chipmunk burrow. In common with the other immature accipiters, its back was dark brown and its vest striped in brown and white. It held its smooth head, with strongly hooked beak, fiery golden eye, and white "eyebrow," as proudly as a king, but gradually it began to shift uncertainly from one big yellow foot to the other as the jays dived and circled and said unpleasant things in bird language. It held its ground until I stepped outside and sent it elsewhere for breakfast.

Ranging down in size from the goshawk are the fourteen- to twenty-inch Cooper's hawk and the ten- to fourteen-inch sharp-shinned hawk. When mature, both have slate-gray backs and tails and lack the white eyeline that identifies the goshawk. Their white breasts are barred with rusty-red instead of the goshawk's gray. At the middle of the size range they are hard to separate, although the Cooper's tailtip is rounded and the sharp-shin's is notched.

A lifetime observer of birds could probably separate these hawks at a glance, as one recognizes close friends at a distance. In eight years here, I have come to know some of our birds so well that a passing shadow identifies them. Others I recognize when I see them clearly. Still others remain disembodied voices. There are many that I have glimpsed or briefly heard; I *think* I have placed them, but I am not sure. These are not yet on my life list. I prefer a small, sure list to a large one of glimpse-and-guess species.

There was no doubt of the identity of the sharp-shinned hawk that shot across the yard to perch on a stump last spring. It was not much larger than a robin and the sun brought out blue lights in its slaty head and back and tail and brightened its red-barred breast. It flashed from the stump to the doorstep in a way that justified its name of "little blue darter." It clutched the step edge with its yellow feet and bent down, trying to peer under the woodshed, first with one eye and then the other. I could hear, from beneath the shed, the tremulous *squeak-squeak* of a chipmunk that had narrowly escaped disaster. The hawk, effectively separated from the chippy, looked almost wistful. Then, with a blur of wings, it rose, whirled, dipped, and lifted from the ground only a few feet from Ade with another chipmunk in its talons.

When the sharp-shin arrived, our chipmunk population was so large and hungry that the ground vegetation was in danger. I saw nineteen chippies nipping off new green sprouts at one time. The little hawk hunted for four weeks. I found a gray jay's tail feathers, and a warbler ceased to buzz in the trees, but the chipmunk population was reduced to a level safe for the growing things and for the vegetarians that would feed on them. The hawk drifted on, leaving its gift of vegetables and flowers.

In late winter I saw, against the white sky glare, the silhouette of a falcon, made of straight lines and obtuse angles, with triangular, slender, pointed wings, and a long slim tail. A shape without detail, it circled slowly, not very high. I held my breath.

Was this the American representative of the peregrine falcons, those fabulous wandering hunters that drop on their prey like bolts from Olympus and have been timed at one hundred and eighty miles per hour? Even when called by so prosaic a name as "duck hawk" these birds move in an aura of romance, sprung not only from the days when falconry was a sport of European nobles, but also from their extraordinary swiftness, strength, and courage.

The bird broke out of its pattern and swooped to the top of a smallish cedar twenty feet from where I stood. It looked at me as though it might never have seen a human before, and well it might not have, because this was not the peregrine, but the rarest of all visitors from the Arctic, the king of the royal family of raptors, the white gyrfalcon.

It had honored our clearing with its presence because even a royal bird must preen wind-ruffled feathers. Carefully it tended its wings and breast, then fluffed and settled and posed on the treetop. Its broad, heavy breast testified to its great flight power. Its feathers were like fresh snow. Even its beak and feet were pale. Its pupils were circles of black fire, and the dark dotted lines on its back and wings, the chevron marks on its primaries, were like ermine tails on the coronation robes of nobility.

It stayed long enough for me to store the memory of its beauty, then lanced upward and turned straight into the north, toward the open spaces over which there is room for a king to fly, toward the rugged northern cliffs of Greenland or Ungava, where it would rear its young by the light of the midnight sun.

A pair of broad-winged hawks nest somewhere between our two cabins, but I have had no sight of them so clear as that of the accidental gyrfalcon. They reared three young last summer and Ade called me to see the family's broadwinged, fan-tailed silhouettes overhead. Their high-pitched whistling brings consternation to the feeding yard, as only the gray jays have their own chosen paths, through and un-

der concealing branches, by which they avoid all hunting birds and arrive safely on the woodshed roof for food. The broad-wings have never hunted within sight of the house and seem even to avoid other birds. This bears out the sometimes questioned statement that these and other hawks of the Buteo family are enormously valuable because they feed on rodents and large insects by preference.

I have seen one other Buteo here, and that for a few minutes only, while it sailed high across the clearing. It was a red-tailed hawk, and I recognized it even before, banking, it poured the sunlight from the bronze upper surface of its tail. For a red-tailed hawk taught me a lesson, bitter but without price.

During one of my childhood summers a family of five red-tails lived in the belfry of a church near my home. I used to watch one of them every morning as it soared above the branches of a great elm that had survived from Ohio's primeval forest. I had been told about hawks—wicked birds that killed robins and stole chickens and generally were a mistake of the Almighty, according to my adult informants. But I thought the hawk beautiful and, although I was too young to understand, I found it encouraging and inspiring.

Then a neighbor's grandson bragged about his ability with an air gun. My father, wisely deciding that the familiar is not dangerous, had taught me to shoot when I was hardly big enough to rest his .22 rifle on a support and aim it. The next morning I took the .22, called the boy to witness, and shot the hawk out of the sky.

Much puffed up, I took the hawk's head to the proper

county office and collected a $2.00 bounty. This, added to the praises of my parents, made me quite insufferable.

The next morning I looked from the window. The emptiness of the sky was a shock. I had forgotten that the hawk was gone. Nothing was left but a pile of bloody feathers and two dirty dollar bills. I hated myself, the rifle, the money, and struggled painfully to grasp the meaning of my unhappiness. Slowly I realized that out of selfishness I had destroyed something of great beauty that had brought me only delight. Being sorry could not help. Nothing could help. A killed thing is forever dead.

The Resident Birds

THE UBIQUITOUS English sparrows have not yet appeared around our cabins. If and when they do, we will have no fears for our resident birds, for birds native to such a climate as this are hardy, adaptable, and aggressive. They fly with equal briskness through the sweltering oppressiveness of a ninety-degree day or the paralyzing cold of a thirty-below wind. They are quick to discover and exploit any source of food, and even work together against predators. Among them, the blue jay stands out, both in appearance and importance.

Against the dark beauty of the forest, the blue jay is jewel-colored. The slate-blue of its head and back is washed with amethyst and the wings and tail shade from sapphire to turquoise, with bars of black and insets of white. There is a narrow black collar that extends gracefully from a point behind the crest to accent the soft gray breast. A black marking above the strong beak gives a determined expression to the face. When a blue jay perches on a snow-cushioned branch above our door, crouched so that its puffed-out feathers form a muff for its feet, with its crest laid back and its head cocked to one side, asking for food in its best imitation of the gray jay's soft whistling, its appeal matches that of any bird I know.

Blue color in bird feathers is not produced by a pigment, as are gray, black, brown, red, and yellow, but by the structure of the feather. Light enters through a horny outer layer, passes through peculiarly shaped cells with thick walls penetrated by minute air inlets, and reflects from a grayish-brown pigment beneath the cells which absorbs all light but blue. If the horny outer layer is yellow, the feather will look green. If a feather's cells are destroyed by crushing, or lose their air from soaking in water, it will look grayish-brown. The color that gives such charm to the blue jay and other blue-colored birds is as much an optical illusion as the rainbow, but it is no less real to the eye.

It is ridiculous to call a blue jay bad because it eats eggs. This is instinctive and a natural control of bird populations. The balance of nature does not necessarily agree with individual human likes and dislikes.

It is equally absurd to call them bullies or gluttons. They are about a foot long, and need plenty to eat. That they compete vigorously at feeders is to be expected. They carry away more food than they can possibly consume immediately. Five jays, making repeated trips to our yard, once removed four pounds of corn without stopping. If I throw broken grahams to them, they carry the food into the trees, sometimes tucking it into branch crotches within my sight, and return immediately for more. This food, if not eaten by the jays, will be found and used by other creatures.

Judgment of the behavior of wild things by human standards is true anthropomorphism. One must remember that both wild things and men are animals, but wild things are not people—nor are they stones. They are guided much by instinct and little by learning, and human standards of behavior and esthetics do not apply to them in any way. It is thus completely false representation to seriously say that any wild thing is cruel, immoral, kind, or the like.

The attempt to explain animal behavior in human terms

is not necessarily anthropomorphism, because we have no terms except human ones to apply to other species. For example, in woods where I am a stranger, I may come on a young squirrel walking on a log. If I freeze, the squirrel stretches its neck toward me and moves nearer in little jumps, eyes wide, nose quivering, every muscle tense. Its tail may jerk and its mouth hang partly open. Then a blue jay calls and the squirrel squeaks and scampers out of sight. If I say that the squirrel inspects me with caution and curiosity, then becomes alarmed and runs away, I am giving my impressions of its mental and emotional reactions, based on what I see. If I say it hurries home to tell mother about the giant it met in the forest, I am starting a fairy tale; any such passage in a book of fact has been introduced for its charm or humor. If I say that a blue jay sees me and screams an alarm, which incidentally sends the squirrel scurrying, I am giving my impression of the jay's feelings, as well as reporting a common happening.

Blue jays often move in small groups, sometimes silently, sometimes twittering as though conversing. But let any source of danger appear—a predatory bird or mammal or a strange human—and the jays' raucous cries send most of the forest denizens to cover. Small birds disappear into foliage, small mammals flatten and freeze, deer bound away, and I have even seen an adult black bear take off at full speed. I say strange humans, because the jays that frequent our yard set up no alarm when Ade or I move through the immediately surrounding woods. They have even attracted our attention as though seeking help.

There was the morning when the blue jays gathered, screaming, in the trees by the door. A female downy woodpecker was lying in an open space on the snow, uninjured but paralyzed and quivering from shock. When I picked her up, the jays flew about their other business. After a short rest in a dark box, covered with wire mesh so that she could not fly out and hurt herself, the woodpecker revived and clung upside down on the wire. I took her outside and reversed the wire so that she could fly free. Crossing the snow where she had lain were the tracks of a weasel that had passed while the bird was in the cabin.

On another morning, I was writing in an outbuilding. Suddenly there were jays all around the windows, calling, lighting on the sills, behaving in an agitated manner. I went out and the birds flew ahead, then stopped to repeat their screaming and darting around me. I followed them across the feeding yard, which was empty of wild life, and into some dense woods where dead trees lie jack-strawed. The jays began to dart and circle around a tall spruce. On a high stub branch a goshawk perched, its beak open and its wings lifted defensively. When I approached the tree, it shot away, slanting its wings and timing their beat to avoid branches in its path. By the time I had picked my way back, the yard was full of birds and squirrels, and the jays were arguing normally with the hairy woodpeckers over the suet.

The gray jays do not travel through the trees in noisy flocks as the blue jays do, but move quietly, sometimes singly, usually in families. They belong exclusively to the northern coniferous forests and were formerly called Canada (not Canadian) jays. They have many nicknames: meat birds, because they are fond of meat, fresh or carrion; moose birds, because they frequent the big forests that also harbor moose; camp robbers, because they quickly learn that camps are a fine source of food and sometimes steal from some outdoorsman's plate; and whiskey jacks, which may come from an Indian word that might be spelled *wiskijon*, or from the Chippewa name, *quing-guish-ee*.

Their colors are the soft grays of clouds that bring gentle rain. On warm days their feathers lie flat and they have a sleek, long-legged appearance. When bitter cold comes, their plumage puffs until even their heads look larger and they seem to be wearing plush knickerbockers. The dark-gray back, tail, and wings show variable touches of light gray or white. These markings give one of our regular visitors the appearance of wearing an old-fashioned tail coat with back-button trim. The underparts are lighter and the forehead is white, in sharp contrast to the black nape. Their large dark eyes are pleasantly inquiring.

These birds are often said to be noisy, but the only loud call we hear is a ratchety sound, used when the birds are annoyed or are threatening some enemy. Their usual vocabulary is made up of soft whistles, *wheets* and *whoots*, and sometimes melodious trilling warbles.

Two pairs nest near us, one in the woods between the cabins, the other a short distance to the east, each producing two broods of one to three yearly. During mild winter days they mingle freely in the feeding yard, but during frigid weather and while their young are immature they are very competitive, diving at birds not of their own group and striking with their wings. Within the families, when all are adults, feeding is a matter of first-come, first-served. This makes more difficult the situation of a bird that has lost its left foot, yet has survived the perils of the woods for two years.

It tried to hang to the suet cages, maintaining leverage by flapping its wings, but this quickly exhausted it. Because it is off balance on the ground and slow to rise, having only its wings for lift, it has learned extreme caution. It usually comes to the yard alone and perches near the door, holding its single leg diagonally underneath as it squats low, touching wingtips to the branch for balance. If one of us does not see it and come out, it flops onto a windowsill to attract our attention, then returns to its former perch from which it can easily launch itself. We throw out pieces of suet and graham cracker from which it selects, sometimes picking up and discarding several pieces until it finds the largest, which it carries away. If blue jays are attracted down from the trees by the shower of food, they wait on the sidelines until the handicapped bird has eaten.

This is unlike the blue jay's commonly observed aggressive tactics, and points out how much is to be learned about animal behavior and perception. Many excellent controlled experiments have been made, but often these show us only how well the animal being tested can do human things; the experiments do not measure the native abilities of the animal that man lacks, has in unimportant measure, or in different form. Clues to the animal's ways and intelligence in its own sphere may lie in the exceptions to its usual behavior.

The earliest gray jay brood flies through our woods with their parents during the first week of May, but they may appear earlier in less rigorous climates. They are full-size, and have very dark-gray juvenile plumage. They are distinguished sharply from the adults by their almost black heads. Because their pink mouth-corners are tender, I offer moistened grahams or soft bread and the youngsters quickly follow their parents to my hands. On sharp spring days they sometimes sit down for a short rest with their chilly feet tucked snugly between feathers and palm.

These birds travel widely in all weathers looking for food. They eat omnivorously. We have seen them eating slightly soured canned carrots and picking our Christmas turkey bones alternately as though to balance their meal. A deer, injured by wolves, rested within sight of our windows some winters ago and the gray jays inspected it often. When we saw them begin to eat the frozen blood, we knew the deer was dead and we could remove the carcass.

I was walking along the side road a mile from the house one day and was thrilled when the pair of gray jays that nest between the houses glided down to me from the tiptop of a black spruce. Their flight is beautiful and, as they sail with feathers spread and perfectly motionless, you can see every detail as clearly as in a high-speed photograph.

Gray jays nest very early when the snow is still on the ground and temperatures here may drop below zero. Their nests are so well concealed that they are seldom seen, even when the approximate location is known. The pair that I met on the road rear their families within a close-grown clump of black spruce and pine, the lowest branches of which are fifty feet from the ground. These birds show great trust in Ade and me; when we walk through the woods, they will fly to us from the nest that holds their eggs or nestlings. Sometimes the chickadees are said to be the friendliest birds in the north, but our vote goes to the gray jays.

The flock of chickadees that winters over with us numbers in one year as few as six, in another as many as thirty, but always has both black- and brown-capped individuals. The black-capped birds are more numerous. Their inky-velvet feather caps, their prominent black bibs, their light-gray backs and buffy breasts are as familiar as their cheerful repetition of their name. The brown-capped, just as charming but more reserved, are slightly longer, with brown-velvet caps, dark brownish-gray backs and tails, grayish breasts with rich light-chestnut sides, and the black bib is not so noticeable in their darker coloring. They speak more softly than their cousins and lisp, "Zhicka-zhee-zhee-zhee."

The chickadees' small wings brace them as they sink into tremendous drifts of snow, all of three inches deep. They fly through gales of subzero air, sometimes wobbling on course, but never losing way. Their energy is enormous, and so is their need for food. We once saw a black-cap lying in the snow beneath the suet feeder. Although these little birds light all over us when we go outside, especially preferring to dig their claws into Ade's thick wool shirts, we have never touched them, and now the exhausted bird hopped away when I tried to pick it up. Ade warmed suet lightly, chopped it fine, stood as close as the bird would tolerate, and tossed the food to it. It picked a little suet, rested, then took a bit more. Fifteen minutes later it had recovered enough to select a good-sized piece and fly away home with it.

Home was probably one of the many cavities that exist in the trees, living and dead, and the stumps of this old forest. Chickadees will also settle in nest-boxes to lay their six to eight half-inch-long eggs. Casualties here seem to be fairly high, for I have seen no more than three fledglings with a parent.

Most birds might be said to have homes, in the sense of nests, only when they are tending eggs and young. At other times they roost where they feel protected, most of the land birds tucking the beak under the shoulder feathers to secure the neck. When the legs are bent, a tendon tightens and curls the toes around the perch. There is a locking device in the toes which prevents slippage of the tendon, so that the sleeping bird does not fall.

The brown-caps are rather retiring, but the black-caps will tackle anything, size being no deterrent. We have seen them join the jays to dive at an intruding German shepherd dog and, alone, they will pester a weasel into withdrawing—if they are not careless and get within reach of its leap. One December afternoon, Peter, our whitetail buck of gentle memory, came into the yard to have some corn. An outraged red squirrel ran up one of his hind legs and, while he was shaking off this nuisance, a black-capped chickadee lighted on a tine of his thirteen-point antlers. *"DEE-DEE-DEE!"* shouted this valiant defender of the corn pile in a deep baritone. Peter, raising his head with a look of mild surprise, rotated his ear, so that the chickadee looked as though it were about to fall into a ship's ventilating funnel. At the next loud cheeping, Peter hastily withdrew his sensitive ear and shook his head, dislodging one ready-to-drop antler that fell into the snow with the chickadee still clinging and protesting loudly.

Slim and trim in summer, fluffed into feather balls in winter, the chickadees give an impression of constant movement—perching briefly on twigs and stems, flitting about with blurred wings, snatching a sunflower seed from a feeder, sitting on a branch with the seed held tight by the feet, while the short beak taps and taps, and finally releases the kernel. They, and the red-breasted nuthatches that sometimes accompany them through the woods, are the butterflies of wintertime.

These nuthatches also wear black caps, but have a black line through the eye and no bib. Their backs and wings are blue-gray, and their breasts are flushed with crimson. Their tails are stubby and their beaks long, in contrast to the chickadees' long tails and tiny beaks. (The white-breasted nuthatch—which prefers deciduous woods, while the red-breasted is at home in both deciduous and coniferous—lacks the black eyeline and bright breast color. I have seen none here.)

The name nuthatch, which would be more accurate in meaning if spelled "nuthack," comes from Middle English *notehach* or *nuthake*. Changing the guttural *ch* in *hach* to the sound in "chance" led to the modern verb, hatch, meaning to mark with repeated strokes as in shading a drawing, and to the noun, hatchet. Thus, a nuthatch hacks for its food by repeated strokes of its beak.

One hears the nuthatches' piping, which reminds Ade of his fourth-grade teacher's pitch pipe, long before one sees them coming through the woods. They move along slowly, hopping up and down and sideways on the tree trunks, hunting out insects and their eggs in the chinks of bark, hanging head down as often as head up.

Last July a parent brought two fledglings to our suet

feeders, possibly from a nesting cavity in a tree. One of the youngsters, watching the chickadees take graham cracker from my hand, decided to try this. It came to the eave over the door, hung upside down on the edge, and dropped to my fingers under the shelter of the roof edge. It did not cling, as the chickadees do, but hopped on the palm of the hand as lightly as one of its own feathers. Its tricky way of asking for a handout is unique among our birds. This family stayed here all winter, along with half a dozen other nuthatches. As our nuthatch population is fairly stable throughout the year, we assume that they are residents, although some red-breasted nuthatches migrate from the more northern part of their range.

Because of their persistent and year-round search for insects, the woodpeckers are of inestimable value to the forest. The hairies and downies are most commonly seen here.

Both of these woodpeckers are black and white, the only reliable plumage difference being in the outer feathers of the tail. Those of the hairy are pure white, and those of the downy are lightly barred with black—but this distinction is hard to see when they are far away or moving, and is hidden when they are at rest. They are most easily distinguished by size and head shape. The hairy is as long as a robin, although more slender, with a large beak that gives the head an elongated appearance. The largest downy is not so large as the smallest hairy, and its head is round, with a short, trim beak. In profile, the downy's head is not unlike that of an outsized chickadee. Because of its short face, the downy has a young look, and it is sometimes mistaken for a fledgling hairy. The males have patches of brilliant, almost fluorescent, red on the back of the head. The male downies that visit our feeders have one centered red patch, and the male hairies two red patches, placed at the sides of a vertical black bar. This separation does not always occur, and so is not a reliable identification mark.

These woodpeckers announce themselves by drumming or cackling. The hairy's call sustains a single note; the downy's, which we hear less often, decreases in pitch.

They feed similarly, moving up a tree bole in a series of jerky hops, picking eggs from under bark scales, now leaning back, braced against the trunk by stiff tail feathers, pecking at a small hole and extending a barbed and sticky tongue inside to bring out the hole's wood-boring inmate. Then they fly in undulating swoops on stiff-feathered, noisy wings, to another tree, where the hunting sequence is repeated. A female downy once discovered some nail holes in one of the posts that support our woodshed roof. She came back day after day to chip at the wood in a fruitless attempt to reach the nonexistent grubs. If there are holes in wood, there *must* be food! Ade finally covered the post with metal so that she might use her time more profitably and not cut the post to the breaking point.

These woodpeckers were very shy when we first put out our suet, peeking at us from behind tree trunks with an air of pretending that they were invisible. Now that they are used to us, they are our alarm clocks. If the feeders are not out when they feel it is time, they drum on the logs and the windowsill to remind us of our duties.

In 1960, I was roundly criticized by an ornithologist who stated flatly that woodpeckers drum only during their spring mating season, when the drumming is a substitute for song. Since then I have taken particular note of our woodpeckers' drumming. All of them drum in the spring and sporadically through the remaining warm season. The hairies and downies not only rat-a-tat at the windows, but occasionally drum elsewhere in winter. One pileated woodpecker drums at any time of the year, often on a hollow tree twenty feet from the living-room window. There can be no possibility of a mistake when bird, tree, and calendar are all within my sight.

When Ade takes the feeders out, the hairies and downies fly ahead of him, wait on low branches, and drop down before his hand is out of the way. When they have eased the pangs of hunger, they settle against the tree trunk, leaning on their tails, with heads lowered and eyes closed, dozing, perhaps as an aid to digestion. One male hairy, when he rouses from his after-breakfast nap, drops to the ground and, turning his head sideways, uses his beak to scoop up an unwoodpeckerlike ration of cracked corn.

When the hairies and downies investigate the bark of the three-foot-thick white pine at the edge of the clearing, they seem dwarfed, but not so the magnificent pileated woodpecker. This bird, sometimes twenty inches long, has a splendid scarlet crest that stands out against the darkness of the pine bark. The male's head is further brightened by a streak of red, extending backward from the beak; the female lacks this streak and the front portion of her crest is black. Black-and-white stripings extend from the beak down the bird's sides. When it clings to a tree trunk searching for food, it usually looks all black except for the head and neck, but when it preens, the white underside of the upper wing shows. In flight, it presents a beautiful black-and-white under-pattern.

The pileated belongs to big timber, whose thick trunks give it nesting sites as well as supplying the carpenter ants that are its favorite food. The felling of the virgin pines reduced the bird's numbers until it was rare here. Now, although it is not common and may never be so again, it has adjusted to the larger second-growth timber, and is found in small and scattered numbers. The nest-hole entrance is an upright rectangle with rounded corners, usually high in a tree. A particularly fine one, abandoned now, is only six feet above the ground in a thick and healthy cedar across the road from our gate.

This woodpecker is often heard at a distance, making a loud *kak-kak-kak-kak* or whacking so sturdily that its blows might be mistaken for human chopping. This is especially true on windy days, when sound travels far and is distorted. The pileated whacks away at stumps, cutting first from one side, then from the other. In rotten wood, the chips are small splinters, such as one might dig off with a fingernail, but in sounder wood the chips are neatly cut, as with an ax. I have seen thin, wedge-shaped ones almost three inches long and tapering from an inch to a jagged point where the chip broke free. The pileated cuts neat round holes into tree roots to get at ants burrowing upward into the trunk. There are several of these holes in an aspen near the log cabin. The woodpecker's activities stopped when the ants were devoured, and the tree, now free of insects, has years to live. In a balsam the cutting has gone on until the bottom two feet of the trunk is riddled. The chips are spongy, and it is plain that this tree is badly infested and has been past saving for some time.

One male pileated makes a tour of the yard at irregular intervals. Although he investigates trees within a dozen feet of the house, he has not approached nearer than fifty feet when we were outside. He starts at the big pine and follows a regular route from trunk to trunk, tapping and examining. If a tree passes his inspection, we know that it is insect-free and sound. It will withstand any ordinary wind, and we do not have to consider whether or not it may fall on the roof. If the pileated starts to work on a tree's roots, we wait. He will either save it or warn us that we must cut it, should it be dangerously located and unsound. He is not only a joy to see, but a very useful bird.

The arctic three-toed, now officially called the black-backed three-toed, and the American three-toed woodpeckers are remarkable for yellow feather patches on the top of the males' heads. Otherwise, three-toed woodpeckers are black and white. The arctic has a solid-black back; the American, black-and-white ladder markings down the back. A visiting friend believed that she sighted one or the other of these birds near our summer house in July, 1961, but neither Ade nor I have seen one. They may well be here; they usually peel bark quietly to find their dinner, and so do not announce their presence like the more noisy of their family, but Dr. O. S. Pettingill, Director of Cornell University's Laboratory of Ornithology, tells me that the arctic three-toed does hammer.

There is no mistaking the *cronk-cronk-cronk* of the raven. One individual, Mr. E. A. Poe, circles the house every winter morning, then flaps away croaking as though disappointed that we are still alive. Come spring, he cheers up and sings rather charmingly, in an exotic, disconnected way, to Annabelle Lee, the feathered lady of his choice. They have nested nearby, probably at the top of a secluded conifer, for several years, but we have seen no young.

Contrary to common misconception, a raven is not a very large bird. It is actually only one-fourth longer than a crow, although it is much heavier. When flying, they can be told apart by the shape of the tail, the raven's wider in the middle than at either end (called wedge-shaped), and the crow's straight across the tip. The raven's wings are flat when soaring, and the crow's slant upward from the body. When perched, the crow looks smooth and neat, while the feathers of the raven's throat hang loose and ragged.

Our ravens are so quiet and reserved in summer that some people refuse to believe that they live in this woods. But we hear them occasionally in warm weather, and they are very conspicuous after the crows have gone south. In winter Ade has heard their noisy croaking as they clean up the remnants of a wolf-killed deer. A dozen may rise into the air if he walks near their feast. I may not envy them their choice of food, but I would, if I could, reward them for their part in keeping corruption from the waters. However, our Mr. E. A. Poe does not seem to like people and we have rarely been able to attract him with food.

Last spring the barking of a fisherman's dog led me to a ruffed grouse's nest in the grass under a honeysuckle bush. The mother had apparently tried to decoy the dog away, but was rising into the bush, leaving some of her black-banded tail feathers behind. The dog plunged into the nest as it jumped away, and crushed six of the eleven eggs.

Shortly after, Ade was shown a grouse chick in a cage in the broiling sun. It had been "rescued" by well-meaning summer residents who hoped to rear a young thing they believed lost. Ade moved the cage into the shade, where the fainting chick revived, but it died in convulsions after being fed flies killed by insecticide.

Two days later I took another chick away from a cat that was playing with it. The delicate baby bird had a broken wing and leg, and several tooth punctures in its back. It was beyond hope of recovery.

At the end of the week, Ade found two more dead at the edge of the road, apparently a result of that morning's spraying with brush-killer.

Within two weeks, all but one of Mrs. Grouse's clutch of eggs had come to grief. Ignorance and thoughtlessness lie behind such tragedies in the wilds.

It is not kindness to let dogs and cats run free in wilderness like this. Not only do they take a large toll of helpless young wild things, but they themselves are in danger. In the summer of 1953, a female beagle was eaten by a wolf in this county. Every year, dogs and bears conflict. An encounter with a sick wild creature may pass on rabies, distemper, or parasites. And, if a tame animal is lost and cannot be found before its master must leave for home, it faces a miserable death from winter cold and hunger.

It is a mistake to pick up young birds or animals, assuming that because they are alone they have been abandoned. Wild mothers rarely separate permanently from their young until they are able to care for themselves, and human-reared wild things are lonely and confused when returned to their native environment, as well as being handicapped by the lack of training that only their own kind can give.

Roadsides, sprayed with herbicide to destroy weeds and brush, are lined with dead and bare saplings, black-leaved shrubs, and withering evergreen that offend the eye. And, if brush *must* be sprayed, this can be done effectively at a season when the roadsides are not full of fledglings. In areas where wildlife abounds, every precaution should be taken to avoid adding hazard to the many that exist in nature. Selective cutting would assure adequate visibility for motorists and, especially in areas visited by tourists, would more than repay any increased cost by retaining both beauty and wildlife.

In late August I again saw Mrs. Grouse, keeping meticulous watch from a stump in the yard, while her single well-grown chick fed on ripe bunchberries. They were joined by a second mother with four chicks, and flew into the berry-laden branches of the elder outside the living-room window. The mothers snatched a few berries, before taking up guard positions on slightly raised large stones. The young flopped noisily through the branches, stopping in confusion when twigs bent under them making the berries seem to retreat as they approached. Two chicks ventured out far enough to reach the berries—and were promptly dumped by the bending limbs. Looking up, they found that the advance of their siblings had lowered the berries within reach; as the others moved out, those on the ground feasted on the drooping clusters. When the branches could no longer support the second group of birds, they changed places with the first, much as though they had worked the whole thing out as a cooperative venture. When the two undersized families moved on, the elder bush wore its one remaining berry cluster like a topknot.

It should be obvious that Ade and I do not hunt. This is often thought odd, but to us hunting is as anachronistic in today's specialized modern world as a privy in Rockefeller Plaza. Our ancestors hunted to make a living, as do certain of our less-civilized contemporaries, using the meat, hides, and bone of their prey for food, clothing, and equipment. Boys among primitive peoples prove their manhood by bringing in their first kills. But modern stores supply improved products that make the older necessities obsolete. Therefore, when a person established in modern business and society tells me that he hunts because he needs the meat, or carefully explains that he carries a rifle into the woods only so that he can see the birds and their nests (at a time when most birds are gone and the nests are vacant), I find it hard to keep a straight face.

And the lack of knowledge displayed by some modern Nimrods would send Hiawatha shuddering into his wigwam. In this area we have the ruffed grouse, usually the gray phase but sometimes the brown, which may be shot during the fall open season. We also have the spruce grouse, which is protected by law. This male has a black neck and chest, black-and-white belly, and dark-gray, barred upperparts; his head bears touches of red. The female is dark and smoky. They resemble the ruffed grouse only in that they are ground birds of about the same size. Yet ill-informed shooters destroy them, then toss them into the brush when apprised of their mistake. According to *Bird Portraits in Color,* even the red-crested, black-and-white, airborne pileated woodpecker has been killed when confused with the brown, long-beaked, ground-dwelling woodcock! The reason? The pileated is sometimes called the logcock. The shooter, hearing a name that sounded like woodcock, and having no idea what either bird looks like, shot the woodpecker.

Hunters and fishermen would do well to cooperate with Federal and state conservation departments and to contribute to the research of universities and conservation organizations, thus aiding the gathering and dissemination of knowledge about game animals and their environment.

Fair Weather Friends

WHILE THE FROZEN LAKE is still safe for and dotted with ice fishermen, spring flaps into the clearing on black wings, settles in a spruce top, and squawks *"Caw-caw-caw."* The sweet song of the bluebird, that best-loved harbinger in other places, could not please Ade and me more than the raucousness of our pair of crows.

They seem to carry calendars in their heads; each year they arrive between March 15 and 17. They strut over the thaw-pocked snow, surveying the debris that appears as the level drops, picking and choosing this and that as suitable for a crow's spring diet. The red squirrels dispute their right to the area, but are routed when the crows sail low overhead. The winter birds ignore the crows with aplomb.

Soon, we hear the plaintive crying of the herring gulls. They have come back to fish through the first cracks in the ice, at the narrow outlet where the water starts its long trip from our lake to Hudson's Bay.

Then comes the passage of the Canada geese, one of the most thrilling sights in the world. First, I hear their calling,

shrill and sharp and clear. Then I see the wedges, frail as cobwebs trailing through the distance, wavering against infinity, coming nearer and nearer until the thin strands of darkness become lines of individual birds. Arrow-straight, their long necks reach ahead; steady and strong their wings beat, along their unmarked skyway. And then they slip beyond the hills, fade into the mists of the horizon, leaving a human watcher—earthbound and exalted, longing in some unfathomable way to share their freedom but not regretting that this cannot be.

The several races of Canada geese range from two to three and a half feet long. They are easily identified by their white "chin straps," black heads, and long black necks that contrast with their pale chests. Occasionally a few snow geese, white with black wingtips, and blue geese, gray with white heads, have been seen in the flocks passing over here. Sometimes lines or V's of cormorants are mistaken for geese in flyways where both appear, but the cormorants are black and silent, sometimes sailing on motionless wings, which the geese never do.

Canada geese mate for life, and nest in solitude and retirement. Although their nests were once common in the Middle West, and have been found in California and Tennessee, the birds have withdrawn from advancing civilization and now rear their families of five to nine mostly in the far north. The nests are usually in grassy or reedy places near water. The grasses supply edible seeds and greenery, nesting material, and concealment; the water is necessary for the safe landing of the heavy birds. The nest may be warmly lined with pale-gray down and, where ground flooding occurs, may be built in a tree, perhaps using an abandoned osprey nest or the like for a foundation. Wherever the nest is, the gander stands guard, ready to protect his family with his life.

From 1954 through 1958 I watched the geese pass over-

head, calling down that the northern waters were freezing in the fall, and that the ice had broken somewhere in the spring. Always they flew between the northeast and the southwest, where marshlands promised shelter. Now, when drought has diminished such marshes as injudicious draining has left, the geese must change their route or perish for want of places to feed and rest. They no longer come my way. Some day, if the drainage programs have not destroyed all the wetlands, when the long dry years are over, the Canada geese may fly their old skyway. Every spring and fall I will listen and watch—and hope, both for the return of the geese and for the restoration of the marshes.

During the weeks following the crows' arrival there has been more thawing than freezing, more run-off water than rain or snow. Patches of open ground appear and I wait for the next migrants. It is not possible to say which birds will come first, or when, because during the past few years there has been such wide variation in the birds' movements that even the species differ from summer to summer.

Climatic changes brought about by shifting of the jet-stream may have much to do with this. This influential upper-air current that, when we moved here, passed across Canada in a west-to-east direction, is by no means constant. It has swooped, with considerable variation, south-southeast from Alaska, passed west of us and, at a point south of the Great Lakes, swung northeasterly toward New England. Thus, the heavy snows that fall south of it have not dropped on our border region since 1956, but have come down in a deep curve to the south, and the Arctic dryness that lies north of it has moved southward. This has lowered lake and stream levels here, dried wetlands, and brought irregularities in the spring and fall movements of the snow line.

On a morning when the snow is not yet gone, I wake and look sleepily out of the window at a shower of gray leaves. But dead leaves do not blow over the snow of a northern spring! Fully awake, I jump up to greet the slate-colored juncos.

In the long shadows they gather around the scattered corn, plump little birds wearing modest, dark-gray hoods above pale underfeathers. There is an argument about a preferred position and two of them rise, face to face, saying "Tchu-tchu-tchu!" which apparently means "I got here first!" or "Find your own place!" The white outer feathers of their tails flash as they fence in the air. Then the argument is over and they settle to pick up broken corn bits and grind them in their pinkish-gray beaks until only husks remain to be discarded.

Still they gather, fluttering from the trees and hopping into the yard. A chipmunk peeks from the woodpile and approaches the edge of the throng, now more than two hundred birds. Soon, chipmunk and juncos are eating contentedly together. A red squirrel runs down a trunk, complaining bitterly. The flock rises with a roaring of wings and the squirrel begins to stuff in solitary splendor. Gradually the birds return to form a twittering carpet around, and at a safe distance from, the squirrel.

When sunshine reaches into the yard, I can distinguish the almost black hoods of the male juncos from the lighter, brownish color of the females. Some of the birds have pinkish side feathers, but the Oregon junco, with black hood and brown sides and back, is not among them. There are sparrows with the flock. Some have chestnut crowns and gray breasts. Of these, the tree sparrows have a black breast dot like a stickpin; the chipping sparrows have a white line above the eye but no stickpin; the swamp sparrows have no stickpin and their throats are marked with white. I see the streaked breast of a vesper sparrow, easily mistaken for a female junco because of its white outer tail feathers and modest brown pattern. And there is a song sparrow, looking much like a vesper without white tail feathers except for a dark spot in the center of its streaky breast. It moves differently, too, jerking its tail as it flits about.

The next day a yellow-shafted flicker stands above the smaller birds like a scoutmaster minding his lively troop. The flicker is the only woodpecker that migrates here from the south, to nest in a hollow tree. The light-brown head with red patch at the back, the barred brown back, and spotted pale breast topped by a black, crescent-shaped bib, are distinctive. The yellow feathers under wings and tail for which the bird is named show only in flight. We will watch the flicker in summer, as it searches the spill-covered ground under the pines for insects.

Three fox sparrows flash their rufous tails. Larger and with striking black-and-white-striped heads, the white-crowned sparrows feed on the windowsill. Ade suddenly points. "There. What is it?" I look, doubtful. Then the bird hops into clear sunlight—a male snow bunting, resplendent in breeding plumage, almost pure white except for the black feathers of his wings and tail. Song fills the evening air and the liquid notes of the white-throated sparrow rise over all, as matchless in their clear simplicity as the double-toned trills of the hermit thrush.

The migrating flock settled to feed and rest, spreading through the forest after breakfasting with us, returning to finish the day with more corn. I noticed a sudden silence one afternoon and looked out onto an empty yard. Then a pigeon hawk lifted a gray body in its talons. One junco would not reach its nesting grounds. When the little falcon had gone, the others returned, their wings and voices closing over the place where the junco had died as though it had never been.

The junco flock left twelve days after it arrived. Two remained, popping in and out of the brush pile as they prepared to nest. Juncos often build on the ground but this pair chose a location amid the interlocked cedar twigs, well above the water in the ditch underneath, and equidistant from the front, back, and top of the pile. I saw them carrying in bits of grass and fern, snippets of moss, and some breast down shed by the gray jays during their molt. My view of the nest was impaired by the intervening mass of twigs, and later by raspberry canes that grew through the interlacement, but I made no attempt to open up the brush pile, for this would have frightened the birds, lessened their protection, and perhaps damaged the nest by moving its foundation. I glimpsed the five spotted, whitish eggs before the green growth hid the nest entirely. Later the parent birds took raspberries from their "doorstep" supply to the young birds, which perched in the branches of a nearby small tree. I counted three streaky young but, as they moved in partial concealment, there may have been more. The adults mated a second time and produced two more babies.

A pair of white-throated sparrows, too, went house-hunting. They chose a thicket of thimbleberry bushes, sprouting maple, and wood ferns. I made no attempt to find the nest in this close-grown mass. Since these sparrows often build on the ground, I might well have stepped on the nest and destroyed its four or five eggs.

Some of the chipping sparrows nested in close-grown small evergreens. In earlier days, they lined their nests with horsehair. When I saw them struggling to shred away the inner bark of cedar, I offered as a better substitute my longish hair that I had trimmed off that morning. The birds accepted my shorn locks enthusiastically. I regret that I was unable to find one of the nests. It would have been quite an experience to see three or four or five blue-green eggs, encircled with black spots at the largest end, lying snugly on a light-brown carpet of my hair.

The tree sparrows moved on to nest in the tundra beyond the limit of trees, and the fox and white-crowned sparrows flew into the north or west, where they might settle in open woods, thickets, or brushy areas. The swamp sparrows sought the open marshy lands, and the vesper sparrows found nesting places in fields or meadows with brush, or perhaps hedgerows. The song sparrows are more partial to shrubbery and thickets, but I think the dense woods around here discouraged them. They lay two clutches of three to seven eggs, but we saw no young and they were not at our feeders in the fall.

By early May the rotting ice has cracked away from the shore. Our flooded brook, pouring into the crack and spouting from its overflow to submerge the adjacent ice, has made a swimming pond, open to any aquatic guests that do not mind ice water.

On the first really warm day in 1960, Ade and I sat on the boat skid. At our feet, bubbles popped to the surface from the ice still fast on the bottom rocks. Musical tinkles announced the disintegration of the glistening edges of the ice cakes that ringed the half-circle of open water before us. Toward the west, the lake was presenting one of its uncommon scenic effects.

At a right angle to our view, the ice had opened in a wide path that reflected the pearl-gray overcast. The windless air was misty above the open path, with a luminous quality that was not quite transparent. The overcast began to disappear and light brightened gradually over the lake. The rotten ice cakes on our side of the crack floated in a coarse mesh of water; beyond the crack the ice sheet was unbroken. As the high fog dissolved into the warming air, all the open water turned blue—dark as ink in the "pond" at our feet, bright between the ice cakes, shadowy aquamarine in the pathway that had been gray. The mist above the pathway thickened and moved to its farther edge, where an updraft built the vapor into a high wall that hid the scene beyond, an opalescent wall made of many rainbows, twisted and twined together so that their colors blended and no single band of color could be seen.

Through the diaphanous top of this wall of light, eight gulls drifted on sun-gilt wings. They soared on the rising air, dipped and rose and side-slipped in and out of the rainbow mist like phantoms. Then they saw us on the skid, and the golden ghosts turned into hungry birds, circling around and over our heads, calling.

Ade got his coffee-can of soda crackers, kept handy in his nearby work building, and flipped a few onto the still water by the shore.

The birds settled on the ice at the far edge of the "pond" and held a conference in muted squawks. All were herring gulls. Four were adults, with bright-yellow beaks, pink feet, immaculate white bodies, soft-gray mantles, and long, black-tipped wings neatly folded. Two were wearing their first-year plumage, grayish-brown all over, with a dark beak. They might have hatched the year before on the farther shore of our lake. The other two were changing from their lighter second-winter plumage to their adult feathers. The gray of their mantles was splotched with brown, their tails were black-banded, and their beaks were ringed darkly at the tip.

An adult plopped into the "pond" and paddled slowly toward a floating cracker, turning his head from side to side to watch us. Having gulped the cracker, he unfolded his long wings and lifted with grace and dignity. The stir in the water sent a cracker toward a partially submerged ice cake where a first-year youngster was sitting. He rose and walked forward, his webbed feet lifting and disappearing under his body like those of a rolling toy. But the submerged ice was slicker than he knew. His feet skidded forward and he sat down with a splat and a squawl. One of the almost-grown-up birds craned his neck to assess the situa-

tion. Then he strolled forward, passed the fallen bird with
a derisive honk, and captured the cracker. This was too
much. The dark youngster attempted to stand up, skidded
again, and, with a crabbed, sidling motion, went into the
water and swam boldly toward a cracker almost at our feet.
Alarmed calls came from the adults, but junior pressed on.
He secured the cracker. Then, seemingly frightened by
his own daring, tried to crouch on the water but succeeded
only in partially dunking himself. Ade threw another cracker
toward him, and a crow flapped out of the trees, hovered over
the floating tidbit, and snagged it efficiently with one claw.
The outraged gull made a sound halfway between the wail
of a tomcat and the hoot of a factory whistle and shot into
the air after the crow, which promptly took refuge inside
the work building, with which it seemed perfectly familiar.
(No doubt this explains the mysterious disappearance of
certain shiny radio parts, left on a bench inside the open
door.)

Then all the gulls were in the air, circling and diving as
we fed them. I have been annoyed by people who said that
something or other was "simply indescribable"; after watching those gulls, I am more tolerant. I do not believe that any
language has words adequate to portray their fluid motion,
their strength and grace, the accuracy of their bankings and
turnings—their landings, when they hang motionless on
curved wings, then drop, without splash and with wings
folding, onto the water. I wish everyone could watch a
group of them at close hand as Ade and I did that day. There
would be a different feeling about the destruction of their
nesting grounds and the spraying of their seas with oil.

For a short time before flying north, a pair of golden-
crowned kinglets decorate the twigs of a small maple like
Christmas ornaments. Ade, while he is walking for the mail,
sees two hundred robins gathered at the edge of the forest.

Later a male sings in the yard, announcing his occupancy to interlopers. The robin's yellow beak, black-gray head, back, and tail, and rusty-red breast, duller in the female, are everywhere familiar, and its song, *"Cheerily, cheerily, cheer up!"* is very heartening on dull days when many birds are silent. The spotted breasts of the fledglings show their relationship to the thrushes. The nests, built of grass, twigs, leaves, and string, and plastered with mud, are placed in crotches and on limbs of trees, under eaves and in open sheds, so that the four light-blue eggs are often seen and the rearing of the nestlings may be a public affair.

Robins flourish in towns and city parks, and they have withstood the advance of civilization very well until the recent prevalence of spraying for Dutch elm disease with DDT. Many robins have died from eating earthworms that had ingested lethal quantities of the poison. But robins breed over most of North America, and still find places where they may live unmolested. In our thick woods, we cannot determine their dates of arrival and departure, though they may be very regular. The robins arrive every year at Christmastime in the gardens of a retreat house in the southeast, where they feed on the berries of holly and pyracantha.

Some purple grackles tamely pick up crackers the first time we toss them from the open door. These foot-long "blackbirds," with their green-and-purple iridescence and wedge-shaped tails, would find excellent nesting sites in our tall conifers. However, as they lay four to six eggs and I have never seen their young, I doubt that any nests are on our property. They avoid the crows at migration time and it may be that, since both birds are partly carnivorous, the larger crows offer too much competition for food.

In 1959, the crows built their nest forty feet up in a spruce by the brook. The squawls and squeaks of the nestlings encouraged me to brave the beaks and wings of the irate parents long enough to see five heads bobbing above the nest.

This was a very successful nesting, for crows lay only from three to five eggs. Some of our neighbors kept minnow pails in the cool brook pools during the summer and the crows abandoned the too-often disturbed location as soon as the young could fly. They now build in a dense clump of trees where traffic is negligible.

A Harris' sparrow, outstanding with his black bib and face, visited for a few days. He was on his way from the plains west of the Mississippi to limited breeding grounds at the tundra's edge, northwest from Hudson's Bay. The nest, which holds three to five eggs, is on the ground, perhaps sheltered by bushes of Labrador tea. Sometimes this sparrow sang his love song for us, a clear, repeated note much like that of the white-throat, but with an added plaintiveness, as though he were longing for his northern land and his mate.

There is a flash that sits motionless on a branch long enough for me to see that it is a black-and-white warbler. This means that other warblers are here, or soon will be. Our suet cages and compost heaps attract a few, especially when late spring cold delays the insects' rising. One spring a pair of Blackburnian warblers stayed for three weeks, their orange cheeks and throats brightening the shadows. In 1961, myrtle warblers, whose dark breasts and bright-yellow patches give them a dressed-up look, nested somewhere in the evergreens. I saw them in midsummer, feeding their three young, which were perched in a small white spruce. The dull-brown babies were partly hidden, and I could not see the yellow rump patch that all myrtle warblers have, whether in breeding, winter, or immature plumage.

This morning I saw a magnolia warbler that I thought, at first, was one of last year's myrtles; but then I saw that he did not have a yellow crown. He did not spread his tail enough to show the broad white band that is his clearest field mark. This warbler is often called the black-and-yellow, yet I see the head and back as faintly blue and the wings as brownish.

There are some sixty species of wood warblers in North America and more than twenty of them may flit through our trees like frolicsome flowers at migration time, or nest in this woods or in other types of terrain nearby. The family name is misleading. Although the ovenbird, the yellow-breasted chat, and the northern and Louisiana water thrushes sing charmingly, most of the warblers have thin or unmusical songs. The black-and-white's notes are almost above the range of human hearing, and the worm-eating and blue-winged, neither of which come this far north, sound respectively like distant riveting and contented snores. However, the birds' cheerfulness, beauty, and appetite for insects override any deficiency in song.

In June, the woods—which has been almost silent while the resident birds are quietly nesting—fills with voices. Chirps and warbles, trills and whistles, deep-throated buzzes and high sweet notes mingle and separate in a fascinating, and bewildering, symphony. I am most puzzled when flocks of wanderers are present because I hear these birds infrequently and their chorus often adds the calls of several species.

One of the most useful helps to discovering which birds are hidden in thick cover is listening to and memorizing calls from the fine recordings of bird songs; also, a song heard in the open may be recognized in a recording. However, to the close resemblance of different species' songs, and to the variations from place to place, are added differences in song between like singers—even differences in virtuosity.

A friend once called me to the window of her cabin to hear a melody which I thought came from the talented throat of Billy, her tame canary on the porch. The singer was a common goldfinch, a chunky little fellow, with black headpatch and wings prominent against a bright yellow body, a bird not frequently seen here in summer because it prefers to nest in the shrubbery of more open country. His aria was a series of glorious trills and glides, much more varied and lovely than the "wild canary's" usual song. (A week later my friend covered her canary's cage and sprayed the porch with household DDT. Billy lived three days, drooping and quivering, but trying to sing as long as his small life lasted.)

As the days grow warmer, the thrushes move about on the ground, modest and shy, mild and gentle. A favored summer brings four species, all with brown upperparts and lightly spotted white vests: the rusty-brown veery, the rusty-tailed hermit, and two less brightly colored, the gray-cheeked and the olive-backed (or Swainson's). The last two are most easily told apart by the latter's white eye ring. Once, in the spring, we saw four wood thrushes, with rufous heads and white vests polka-dotted with black. They are strangers to the northern evergreens, and they left the next day for the deciduous woods where they breed. The gray-cheeked thrush also moves on, to build in a bush or on a low limb in the far north. The other three may nest near us: the veery on the ground, the hermit on the ground or in low shrubbery, and the olive-backed in the sheltering thickness of spruce or fir. All of these birds lay four eggs, although the gray-cheeked may lay only three. The eggs are beautifully colored—those of the wood thrush like the robin's, the others greenish-blue, spotted in the cases of the gray-cheeked and the olive-backed.

The hermit thrush has inspired many poets and is often said to be the most beautiful singer among American birds. Perhaps this has drawn the spotlight from the other thrushes, all of which have their moments of glory in song. To me, the evensong of the olive-backed thrush is the most beautiful. As the light dims and tired humans settle to relax, this thrush's song comes from the dusky, whispering woods,

rises from its low opening notes in a spray of melody to die away in a high, thin whistle, as though the song had floated up into the sky. The olive-backed thrush sings for human dreaming.

As soon as the ice has broken, another call rings out, one that has been so long in the forests of the north that it is a prologue to all other northern bird sounds, the call of the common loon. In addition to conversational hoots and honks, there are three definite calls, perhaps more than three, with many variations. So many writers have described these voices as shrieks of maniacal laughter that they are often not recognized by those unfamiliar with them. I once heard a guest at a lodge ask what had been howling in the night. The desk clerk, hired for the summer and not prone to study bird calls, said, "Maybe it was a dog."

One of the loon's cries does resemble shrieks of laughter, but to me it sounds hilarious, rather than mad. Another is a mocking hoot, very effective when repeated machine-gun fashion. Once Ade and I, out in the boat, drifted off the shore of a little island that had been blasted almost bare by the wind. I said that it would be quite a place to live, wouldn't it?—and two loons popped up beside the boat, tootling *"Ho-ho-ho-ho-ho-ho-ho!"* in unison as though to laugh a silly human off the lake.

The third call speaks of solitude and quiet, of darkness on still waters, of a land in the days of its beginning. It drifts on the moist air over the water as three mellow notes, the middle one longer and higher than the other two. A visitor from Honolulu told me that it must be a homesick Islander, calling, again and again, "Oahu." Very late one night, while I sat by an open window, a loon just off shore began to call. The phrase had four notes, slurring smoothly one into the other, and it was rich and full and flowed like honey. From the hills came echoes, *ahu . . . ahu,* and from miles down the lake—for this call has great carrying power—an answer murmured, with its own whispering echoes. Then came a long wail, as full of desolation as a cry from the earth's last living thing. I have never heard anything more sorrowful or more beautiful, nor have I ever felt more alone. Yet the mournful sound did not bring loneliness in its ordinary sense, but an acceptance of the essential and natural isolation of the human spirit. Perhaps this is one of the great lessons to be learned from wild places and their dwellers.

Like our other water birds, the loons are with us only from the ice break-up to the onset of cold weather, so we see them in summer plumage. The beak and neck are black, and the black head looks greenish in sunlight. The throat bears a necklace of fine, vertical, black-and-white striping. The back looks as though covered with small, square, white tiles, laid in a black base, and the sides and wings are black with white dots. The breast is like snow.

The birds often float so low in the water that one sees only a little island of back and the dark head and neck, with a narrow white "collar" below. Then the loon disappears in an expert dive, to pop up a hundred feet away. If a ducklike bird goes under water and stays so long that it seems it must surely have drowned itself, it is very likely to be a loon.

Sometimes they float idly, hooting to themselves or to each other. Sometimes they play boisterous tag, one diving to come up under an unsuspecting cousin and upset him. This leads to great running and splashing along the surface of the water, with low flights and much "laughter," as the upset loon seeks to pass the favor on to another of the group.

A pair of loons nest on our shore, probably very near the water because these birds walk awkwardly on land. A loon lays only two eggs and Ade and I, who feel that there are enough natural disturbances, have not pinpointed the location. We wait for the appearance of the little balls of black

down that sometimes hop onto mother's back and ride across the water.

Some two hundred feet from the loon's summer home, the shore is heaped with boulders. Behind the boulders, there is a cavity where Mrs. Merganser's clutch of eggs—looking much like large, heavy-shelled, creamy hen's eggs—rest on a cushion of soft vegetation and down. The eggs may be covered with a down blanket when mother leaves the nest for a short period. The mergansers also nest in tree cavities, in abandoned buildings, and under tree roots near water, apparently liking shelter from rain and enemies.

The American mergansers are large ducks, about two feet long, some eight inches shorter than the loons. They swim sedately back and forth along our shore, diving for a bit of fish, stepping ashore to preen a disarranged feather. The male is uncrested, with a black-green head, black back, black-and-white wings, and pale sides. The female is gray, with a rusty head and a small crest that looks as though it were blown by the wind. The breasts of both male and female at close view show a delicate peach color.

We first saw a nesting pair in 1955, nervously examining a boat skid that Ade had just completed. A pair has appeared every spring since. They seem perfectly familiar with every cranny and stone, and use the lower part of the skid for preening and drying after a swim. Because they are increasingly indifferent to Ade and me, we think it may be the same pair each year. One bird might, of course, bring a series of mates, but this does not account for the growing tameness of both birds.

In midsummer, I hear a splashing at the water's edge. Then mother swims proudly out from shore, a kite's tail of

sometimes *eighteen* ducklings trailing behind her in a gracefully curved line. If one pokes its beak aside to investigate
some floating curiosity and lags behind, mother, who turns
her head often to watch her brood, gives a sharp, harsh
quawk, and junior hastily closes up the gap in the parade.
As they grow larger, mother brings them out in rough water
and the procession rises and falls over the crests of the
waves, still in perfect formation. In 1961 she had only one
duckling, but disciplined and trained it exactly as she
might a more prolific hatching.

In 1962 a pair of common golden-eyes settled about as
far from the mergansers as the mergansers are from the
loons. These ducks can be mistaken for mergansers, but they
are half a foot shorter than the American species, and have
puffy heads and short, dark beaks, as opposed to the mergansers' slim heads and orange or red beaks. The black-and-
white male golden-eye has a conspicuous white spot below
his eye, and the gray female's head is uncrested and brown.

The golden-eyes' nest is in a split spruce stub, where bits
of down cling to the cracked wood about eight feet from the
ground. Like the merganser, the golden-eye mother may
cover her eggs with a down blanket, sometimes so firmly
woven that it may be handled repeatedly without damage.
After the mother appeared on the lake with nine young
(eight to fifteen eggs are laid), I found pale-green shell
fragments at the foot of the stub. Sometimes the loon, merganser, and golden-eye mothers have their babies out for a
quiet swim at the same time. Drifting a little, diving a little,
they leave gentle ripples on the twilight waters. Most visitors
are interested in the birds and do not disturb them, but occasionally someone in a boat chases a mother with young. A

fast boat, cutting across the path of such a little family, can destroy them, of course.

Two or three pairs of herring gulls nest each summer on a small island across the lake. During the brooding time and until the immature brownish birds are able to fly as well as swim, a gull stands like a sentry, from dawn to dark, on rocks that rise from the water east of the island. Ade and I once saw three young gulls swimming under its watchful eye, and thought we might paddle gently near to see them better. The big white bird on the rock was not to be taken in by such tactics. As we drifted in, it flapped and set up a loud crying. From the scrub trees on the island's height came an answering squawl and a bird soared out to circle. Once it had us located, it swooped low over the boat, defecating with deadly accuracy. Our boat is open. As the bird swooped for another attack, we fled.

In late fall, we explored the deserted island and found two nests, little hollows with mossy sides, brightened by intertwined strips of aluminum foil. These strips were "chaff" used, in varying widths and lengths, by the SAC to jam radar frequencies during sham battles. We had found scatterings and clumps of it on our grounds the previous spring, chilling reminders of deadly practice activities in the peaceful sky of the winter wilderness. It was comforting to see these reminders of the ugly use to which man has put his artificial flight being employed in such a homely, domestic way by those whose inborn powers of flight will never be misused.

We see other ducks and water birds, usually at such a distance that their identity is doubtful. In August of 1960, a male common scoter, all black and decorated with bright orange at the top of his bill, rose from a mixed gathering on the water. He banked and turned past our boat so that we

had a fine view of his underplumage, velvet-black on the
body and silver-black on the wings. This, the only all-black
North American duck, breeds in the Arctic and winters
along the northern Atlantic and Pacific coasts. Small flocks
are occasionally seen on the Great Lakes, but the bird is rare
inland. (The black duck, often called the black or dusky
mallard, is not really black but a smoky, dark brown with
a paler head. There is much variety in its plumage, believed
due to its crossing with the true mallard.) On that same
day we saw twenty-seven loons gathered before migration,
many of them in gray-backed immature plumage, waddling
out of the water and into the shelter of rushes at our approach.

Most sandpipers breed in the Arctic, but the spotted
species broods four to five eggs in a sheltered ground nest
throughout Canada, except north of the tree limit, and the
United States, except in the extreme south and east. Ade and
I have seen only one, teetering on its stiltlike legs along the
lake shore. Up and down wagged the tail, as though the
bird were trying to balance on a slack wire. The teetering—
purpose, if any, unknown—is instinctive; a newhatched
chick flips its fluffy excuse for a tail as soon as it can stand.
The bird's upperparts are brown and, in spring, the white
breast has large, dark spots, whence comes the name.

During late August in 1962, five sandpipers not more
than six inches long chose our shore as a resting place in
their migration from the Arctic to the Tropics. From size
alone they might have been the semipalmated, least, or
western species, but their bills were too short to be the last.
Both the others are brown above, light below, and have
white lines above the eyes. The semipalmated has black legs
and the least, yellow or greenish, but the little birds, hopping about and picking on and between the boulders, hid
this detail in the shadows.

Belted kingfishers make themselves known to me as blue streaks with rattling voices. They look like stubby, grayish, blue jays with oversized heads and beaks and irregular crests. The male has a blue band, or belt, between his white neck and chest, and the female has an added bright chestnut band across her chest. Their five to eight eggs are often laid in an excavated tunnel, but the soil here is so peppered with boulders that it seems more probable that they nest in some natural cavity beneath the rocks.

One afternoon I think I hear a distant plane and go out to see whether it is forestry, private, or perhaps RCMP across the lake. I round the corner to find a ruby-throated hummingbird, poised like a golden-green jewel on buzzing wings before the blossoms of the scarlet runner beans. This gorgeous mite is a female, with a white throat instead of the male's red one. Indifferent to me, she darts from flower to flower, her wings only a blur of vibrating shadow.

I wish that I might have seen her building her nest. It is the size of a large thimble, lined with thistle-down and fastened to a wild-rose stem by cables of spiderweb. Bits of green lichen camouflage it and make it as lovely as the two miniature white eggs inside.

The wings of these fairy birds beat from fifty to seventy-five times a second and in the summer they feed every hour. Their use of energy, and the heat loss from their small bodies, must be very great. Yet these three-and-a-half-inch wonders not only migrate to the southern limits of this country, but cross the Gulf of Mexico. It has been said that they hitchhike on the backs of unsuspecting wild geese, but this ingenious idea, I am told by Dr. Walter Breckenridge, Director of the Minnesota Museum of Natural History, is completely unfounded. It is, however, pretty well authenticated that hummingbirds can fly nonstop over water for four hundred miles, the distance from Yucatán to Louisiana.

This seems more astonishing than that they might steal a ride to and from their winter home.

When the longest day of the year has passed, we watch to see how successful our nesters have been. Here is a pair of juncos with two streaked youngsters. The chipping sparrow mothers patiently follow the paths, picking up corn and stuffing the gaping mouths of their children, while the children practice picking for themselves. The young crows look and strut like their parents, only their squeaky, babyish voices proclaiming their newness. One day the warblers fill the trees, and in the flock are many yellow-breasted, grayish-backed birds, with and without wingbars. Some of these are adult warblers in fall plumage, and others are dull-feathered young. It takes years of careful field work to enable one to recognize these confusing individuals.

Then one day the thrush does not sing in the evening, and I miss the jaunty cawing of the crows. The loon's lonely voice fades into the distance, and the gulls do not mew in the dawn. It is time to get in feed for our winter birds again.

The influx of summer birds has varied greatly here during the last years and Ade and I watch anxiously, for we feel that the number and variety of these visitors, as well as their breeding success, may be some index to how the species are faring in more settled parts of the country where poisonous chemicals and other introduced hazards are more frequent than here. The spring of 1962 brought more migrants than we had seen since 1958. Here, if unseasonable storms, forest fires, or spraying to destroy brush or forest insects do not bring disaster, the birds have an excellent chance.

Birds are not only one of the loveliest gifts to the earth, but also are enormously valuable as controllers of insects and eaters of weed seeds. Their lives are hazardous at best. If we want to keep them, we will supply replacement for the food that grows scarce in civilized places. We will erect boxes to replace nestholes that vanished when wooden fence posts gave way to steel ones. We will leave some trees and bushes for shelter—and to alleviate the monotony of human houses, surrounded by sand and concrete. We will endure a few mosquito bites rather than endanger the birds with poison. In sheltering the birds, we will nourish that vision without which the people perish.

The Waiting Hills

FROM THE LAKE, our shoreline is inhospitable looking. Masses of giant boulders extend from the water to a steep bank, beyond which the trees form a dark wall. From the land, it comes alive, and I go there to rest and refresh myself.

The path from the house is smooth with the brown fallen leaves of the cedars that meet overhead. The gentle air is aromatic with rosins and fresh with the scent of earthy fungus. I sit on a room-size, water-smoothed slab of granite, that thrusts into the lake from the boulders that form our natural breakwater.

From the pile beyond the work building comes the sharp smell of damp sawdust. Wild roses are rooted and flourishing in the crevices of the stony bank. An ancient cedar, its lake-side branches stripped and bleached by storms, leans over the water beside me, and birch leaves above my head are green against the tender blue of the sky. The still air is empty even of the shimmer of insect wings. Cloudless and serene, the heavens pale to the west. The water lies smooth as glass, so smooth that it looks deeper than the sky. A fish flops, and a pair of loons drift along, diving for their supper. Little ripples from their passing lave the ribs of a boulder that has been carved by endless waves into the shape of a skeletal monster. A red squirrel comes down the cedar bole. He chatters softly at me, then disappears without haste. A gull cries faintly in the distance, and I catch a golden flash from its wings in the sunset, before it glides away over the hills across the lake—the peaceful, wooded hills.

The earth rests, and remembers.

Stretched out prone on the slab, I notice that a corner has recently chipped away, revealing unrotted interior, a mass of dimly glittering crystals, pink and black and white. I stretch my hand down to pick up a smooth, dark stone, lying just under the surface of the water. I turn it over—and sit up abruptly.

The stone is drying to a brownish black, and its smoothness is only on one surface. The rest of the fragment is a mass of cells, like those of a sponge, connected by porous, cindery rock walls. It is a piece of scoria, formed when gas bubbles were trapped by the cooling and solidifying of a lava flow. The history of this region—not the man-history, but the earth-history, which has been pieced together from evidence left by the great happenings in the days before man—takes on meaning as I sit on the pink granite, holding the vesicular rock in my hand, while the setting sun spreads the northern sky with light the color of the primeval fire that once burst forth there.

Eons ago, so long ago that the time is meaningless, a great mountain range came into being in the area of the Laurentian hills far to the northeast. The crust of the earth heaved and buckled. Volcanoes belched vast clouds of dust and steam and poured forth glowing rock, while fiery fountains danced on the surface of the lava lakes within their cones. The sides of the volcanoes split to pour rivers of molten rock into the newly sunken valleys and to spread blazing seas across the face of the earth.

While this was happening, or perhaps before, or perhaps

after, for the corridors of time yield only mist-dimmed glimpses of their archaic secrets, enormous flows of molten rock spread under the earth's surface. Spread across two million square miles, under rock and under water, reaching north to what is now Hudson's Bay and southwest to the place where the Great Lakes were to form, spread two miles thick and several miles below the world of daylight, to solidify as the pink granite of the Canadian Shield. This took place in stages and over vast periods of time.

Then, as now the red light of sunset fades, so had the molten rocks cooled and grown dim. The rolling clouds of steam that had risen from the volcanic magmas condensed and fell, only to burst again into vapor as they touched the still-hot rocks beneath. At last there came a day when the rocks were no longer hot and the rain formed into trickles and then into torrents, beginning the erosion of the bare and sterile mountains.

But the terrain was not yet finished. More earthquakes buckled the rock layers, threw up ridge upon ridge of mountains, lifted soaring cliffs sheer into the air. And, through clefts in the rocks, lava rose from the depths and again spilled over the land. When these upheavals and flows finally ceased, the foundation was laid for millennia to come, and the rains went about their patient sculpturing of the earth.

While the rivers brought gravel and sand and silt to the valleys, pools and lakes and seas were forming. And, on some most wonderful day, Creation touched the water—and life came into being. But the land was barren.

Then, somehow, somewhere, a fungus grew and held in its folds the living algae. Clinging to a rock, this humble lichen began to drive away the bareness of the land.

Through the passing ages, life forms came and went. The land burned under tropical suns and cooled as the ice sheets crept south from the fields at the pole. Across this place they came, their two-mile-high fronts crushing forests, grinding down mountains, scooping out valleys, pushing surviving living things ahead of them, and carrying the topsoil away to the southerly plains. So great was the weight of the ice that the crust of the earth bent under it, bent and rippled and was depressed hundreds of feet. Then the ice melted and receded, the earth began to rise as the great burden was taken from it, and life moved northward across the lands and waters. When the last glacial sheet was gone, the new-made bed of the lake now in front of me was filled with the melting ice, and the bare-scraped bones of the earth and the slime of the water began to give birth to the forest that I know.

Now the fires of the land are quiet. Volcanoes have not spread their flaming seas for almost half a billion years. The low hills around me, the black bluff that lifts my southern horizon—from what great heights are they worn down? The peaceful hills—resting, waiting. For life-forms as yet unknown, for lush tropic flowers, for glittering blue glaciers, for new forests, for another gargantuan lift of mountains.

Beside me on a boulder a lichen clings, a gray-green tissue of curly crust, nibbling away at the granite as the first lichens nibbled at other parts of this same rock. How long ago? No man knows; no man can ever *know*. But some of the granite of the Canadian Shield looks back beyond two and a half billion years. The slab under me is part of that Shield. The piece of scoria in my hand may have been part of the last lava flow; it may have been brought by the glaciers or been here when the pink granite flowed forth.

From the boiling rocks has come all of earth's past. From their dust will come all of its future. I hold eternity in my hand. . . .

. . . And I am awed by the realization that, insofar as the immediate future is concerned, this is quite literally true of civilized man.

In the days when our forebears competed with the cave bear for shelter and with the dire wolf for food, man was as much a part of and as wholly at the mercy of the whims of nature as any other animal. Now he has learned to control his environment, and this is leading him to lose sight of activity that is not man-induced or man-controlled. He thinks of himself as a creator instead of a user, and this delusion is robbing him, not only of his natural heritage, but perhaps of his future.

Life in the wilderness makes one thing perfectly clear: man cannot combat nature. He may protect himself against its adversities and make the most of its favors, but that is the limit of his power. Whenever the blizzard-tossed heroes of Jack London's stories of the north managed to stay alive, they did it by cooperating with the elements, not by fighting them. If man is to survive on this planet as long as, say, the turtle, he will do it by adapting, not by destroying.

The natural world is dynamic. From the expanding universe to the hair on a baby's head, nothing is the same from now to the next moment. Yet we do not accept this. We try to remodel and stabilize the earth.

In the vastness of time and space, what we do and what we build matters no more than my penciled corrections on a rough-drafted page. But to the generations to come—and whether or not they come depends on us—our actions are all in all.

We stand in ridiculous imitation of King Canute commanding the sea. We write off our failures as negligible and ignore the part played by fortunate timing in many of our successes. When, for any reason and on any scale, we duplicate the reactions by which the sun supplies the earth's energy, let us remember that we may make a mistake that we cannot write off—for *our* atoms, reduced to their component parts, will not reassemble into fingers and pencils. When we poison and bulldoze and pollute, let us remember that we are not the owners of the earth, but its dependents.

Let us look to the earth, to its wealth and beauty, and be proud that we are a part of it. Let us respect it, and time and space, the forces of creation and life itself. As we hold the future in our hands, let us not destroy it.

Bibliography

THE BOOKS AND PERIODICALS listed here are among those that I have found most helpful in my prying into the life, growth, and substrata around me, and that I believe will be interesting and useful to anyone wishing to investigate for himself. All, as far as I know, are in print, and none are expensive. All are applicable over wide areas except four: Minnesota Department of Conservation Bulletins 4 and 5, on wolves and deer in the Superior National Forest and other local areas, and the Minnesota Geological Survey's Bulletins 39 and 41, on local geology and geochronology. These are included because they contain valuable information not readily available elsewhere. All publications listed are nontechnical or contain basic explanatory matter that covers technical terms used, except the geological bulletins. A number of the books I have named are included in *The AAAS Science Book List* (American Association for the Advancement of Science, Washington, D.C.), found in many libraries, which may be used to build a basic library in any field of science.

GENERAL

Buchsbaum, Ralph M., *Animals Without Backbones*, University of Chicago Press, 1948.

Collins, Henry Hill, Jr., *Complete Field Guide to American Wildlife—East, Central, and North*, Harper & Row, 1959.

Coulter, Merle C., *The Story of the Plant Kingdom*, University of Chicago Press, 1935.

Portmann, Adolf, *Animal Camouflage* (translated from the German), Ann Arbor Science Paperback, University of Michigan Press, 1959.

Romer, Alfred S., *Man and the Vertebrates*, University of Chicago Press, 1941.

TREES AND SHRUBS

Harlow, William H., *Trees of the Eastern and Central United States and Canada,* Dover Publications, Inc., 1957.

Petrides, George A., *A Field Guide to Trees and Shrubs*, Houghton Mifflin Company, 1958.

FUNGI

Christensen, Clyde M., *The Molds and Man*, University of Minnesota Press, 1961.

Smith, Alexander H., *Mushroom Hunter's Field Guide*, University of Michigan Press, 1958.

Thomas, William S., *Field Book of Common Mushrooms*, G. P. Putnam's Sons, 1948.

SMALL GROUND PLANTS

Cobb, Boughton, *A Field Guide to the Ferns,* Houghton Mifflin Company, 1956.

Conard, Henry S., *How to Know the Mosses*, Wm. C. Brown Co., 1944.

Housman, Ethel Hinckley, *Beginner's Guide to Wild Flowers,* G. P. Putnam's Sons, 1948.

Mathews, F. Schuyler (rev. by Norman Taylor), *Field Book of American Wildflowers,* G. P. Putnam's Sons, 1955.

Pohl, Richard W., *How to Know the Grasses*, Wm. C. Brown Co., 1954.

Allen, Glover Morrill, *Bats*, Dover Publications, Inc. (reprint of Harvard University Press publication), 1939.

Burt, W. H., and R. P. Grossenheider, *A Field Guide to the Mammals*, Houghton Mifflin Company, 1952.

Cahalane, Victor H., *Mammals of North America*, The Macmillan Company, 1947.

Erickson, Arnold B., *et al*, *The White-Tailed Deer of Minnesota*, Technical Bulletin 5, Minnesota Department of Conservation, 1961.

Stenlund, Milton H., *A Field Study of the Timber Wolf in the Superior National Forest, Minnesota*, Technical Bulletin 4, Minnesota Department of Conservation, 1955.

REPTILES, AMPHIBIANS, INSECTS, AND OTHERS

Breckenridge, Walter J., *Reptiles and Amphibians of Minnesota*, University of Minnesota Press, 1944.

Conant, Roger A., *A Field Guide to the Reptiles and Amphibians*, Houghton Mifflin Company, 1958.

Gertsch, Willis J., *American Spiders*, D. Van Nostrand Co., Inc., 1949.

Graham, Samuel Alexander, *Forest Entomology*, McGraw-Hill Book Company, 1952.

Lutz, Frank E., *Field Book of Insects*, G. P. Putnam's Sons, 1948.

Norman, J. R., *A History of Fishes*, Hill & Wang, Inc., 1958.

BIRDS

Heinroth, Oskar and Katharina, *The Birds* (translated from the German), Ann Arbor Science Paperback, University of Michigan Press, 1958

Peterson, Roger Tory, *A Field Guide to the Birds—Eastern Land and Water Birds,* Houghton Mifflin Company, 1947.

Roberts, Thomas S., *Bird Portraits in Color* (a shortened version of Dr. Roberts' out-of-print *Birds of Minnesota,* 1934, with identical color plates), University of Minnesota Press, 1960.

Sprunt, Alexander, Jr., *North American Birds of Prey,* Harper & Row, 1955.

Records: A Field Guide to the Birdsongs, recorded at Ornithology Laboratory of Cornell University under the direction of Dr. Peter Paul Kellogg and Dr. Arthur A. Allen, in collaboration with Roger Tory Peterson, 1959.

THE EARTH

Dunbar, Carl O., *Historical Geology,* John Wiley & Sons, Inc., 1949.

Goldich, Samuel S., *et al, The Precambrian Geology and Geochronology of Minnesota,* Bulletin 41, Minnesota Geological Survey, University of Minnesota Press, 1961.

Grout, Frank F., Robert P. Sharp, and George M. Schwartz, *The Geology of Cook County, Minnesota,* Bulletin 39, Minnesota Geological Survey, University of Minnesota Press, 1959.

Longwell, Chester R., and Richard Foster Flint, *Introduction to Physical Geology,* John Wiley & Sons, Inc., 1955.

Pough, Frederick H., *A Field Guide to Rocks and Minerals,* Houghton Mifflin Company, 1960.

PERIODICALS

Audubon, National Audubon Society, 1130 Fifth Ave., New York 28, N.Y.

Canadian Audubon, Canadian Audubon Society, 423 Sherburne St., Toronto 5, Ont., Canada.

Frontiers, Academy of Natural Sciences for Philadelphia, 19th St. and The Parkway, Philadelphia 3, Pa.

Living Wilderness, The Wilderness Society, 2144 P St., N.W., Washington 7, D.C.

Minnesota Naturalist, Minnesota Natural History Society, 315 Medical Arts Bldg., Minneapolis 2, Minn.

Natural History, incorporating *Nature Magazine,* American Museum of Natural History, Central Park West at 79th St., New York 24, N.Y.

Index to the Flora and Fauna

HELEN HOOVER (1910-1984) is the author of several books, including *The Years of the Forest*, *A Place in the Woods*, and *The Gift of the Deer*. Before moving to the remote wilderness of northern Minnesota in 1954, she was an accomplished chemist. Many of her books are illustrated by her husband, Adrian. Her writing has been treasured by generations of readers throughout the country.